Dampers

How Viscous Dampers Protect Structures

Douglas P. Taylor & David Lee

Copyright © 2017 Douglas P. Taylor & David Lee
All rights reserved.

No part of this publication may be reproduced or transmitted in any form or by any means, electronic or mechanical, including photography, recording, or any information storage and retrieval system without the prior written consent from the publisher and author, except in the instance of quotes for reviews. No part of this book may be uploaded without the permission of the publisher and author, nor be otherwise circulated in any form of binding or cover other than that in which it is originally published.

The publisher and author acknowledge the trademark status and trademark ownership of all trademarks, service marks and word marks mentioned in this book.

Acknowledgments

Thanks to all the good people at Taylor Devices for making this book and our success possible. Special thanks to MCEER at the State University of New York at Buffalo whose earthquake test lab demonstrated the benefits of added damping numerous times, and to Professor Michael Constantinou and his students who tested, evaluated and developed new concepts for implementing added damping within structures.

Lastly, thanks to David Lee for his help in writing this book.

Introduction

Taylor Devices has been making viscous dampers and shock absorbers since 1956. This is the story of the company. It is also the story of Paul Taylor who founded it, and Doug Taylor who built the company into a technology giant.

The first part of this book tells the story of dampers throughout the ages, leading to their use today in spacecraft, missiles and protection of structures from earthquakes.

Doug Taylor originated the use of viscous dampers for earthquake protection of structures, working in cooperation with his friend Dr. Michael Constantinou of State University of New York, Earthquake Lab. In recognition of this work he was awarded the title Structural Engineer of the Year, even though he is not a structural engineer. As part of this award Doug Taylor was asked to write his autobiography, which forms the second half of this book.

We hope you enjoy this fascinating history of viscous dampers and the people who created this revolutionary technology.

Contents

Introduction v

BUILDINGS: DESIGN FOR DAMPING 1
Abstract 3
Introduction To Dampers: Definitions And Functional Output 4
Fluid Dampers For Building And Bridge Structures 23
Performance Of Fluid Dampers 33
Implementation Of Fluid Dampers 46
Project Examples 58
Conclusions 69
References 71

INTRODUCTION TO PART 2 73
Summary 74
1. Introduction 75
2. The Sandbox Incident 76
3. Growing Up In The 1950s And 1960s 81
4. 'Almost' An Engineer 84
5. The U/B Years 98
6. Entering The Corporate World 111
7. The Space Shuttle 119
8. The Mx Missile 123
9. The Reagan Years 126
10. Seismic Dampers? 129
11. Nceer And The San Bernardino Medical Center Replacement Project .. 132
12. Other Early Seismic Damper Projects 141

13. Damping Concepts For Tall Buildings 149
14. Torre Mayor 151
15. Tall Buildings With Toggle Braces 155
16. The Future 157
References 160

BUILDINGS: DESIGN FOR DAMPING

by

Douglas P. Taylor, President Taylor
Devices, Inc.
90 Taylor Drive
P.O. Box 748
North Tonawanda, NY 14120-0748

ABSTRACT

The end of the Cold War in 1990 heralded a restructuring period for the American military and defense industry. One of the outcomes of this new era was that political and economic change allowed previously restricted technologies to become available to the general public. This conversion of defense technology is typified by highly advanced products and services that suddenly appeared in the marketplace, seemingly out of nowhere. Perhaps the best known of these is the now ubiquitous Internet, which in reality came from 1970's defense technology intended for use by government agencies in the event of nuclear war.

In the civil engineering field, high capacity fluid dampers have transitioned from defense related structures to commercial applications on buildings and bridges subjected to seismic and/or wind storm inputs. Because fluid damping technology was proven thoroughly reliable and robust through decades of Cold War usage, implementation on commercial structures has taken place very quickly.

This presentation is both a broad overview and a guide to implementation; with specific case studies being provided from four of more than 700 major buildings and bridges equipped with fluid dampers by Taylor Devices, Inc., a defense contractor from the Cold War years.

INTRODUCTION TO DAMPERS: DEFINITIONS AND FUNCTIONAL OUTPUT

The concept of damping within a structural system can have different meanings to the various engineering disciplines. To the civil engineer, damping may mean only a reference note on a seismic or wind spectral plot, 5% damped spectra being the most common notation. To the structural engineer, damping means changes in overall stress within a structure subject to shock and vibration, with frequent arguments whether a structure will have 2%, 3%, 4%, but not more than 5% structural damping. On the other hand, mechanical engineers do not necessarily view damping as a benevolent feature, since machines, by definition, are supposed to transmit forces and motions efficiently, without energy losses. Thus the need for damping in a machine often signifies that an engineering design error has been made.

In the classical mechanical engineering text "Vibration Theory and Applications", William Thomson [1] avoids a single, direct definition of damping by offering the following descriptions: "Vibrating systems are all more or less subject to *damping*, because energy is dissipated by friction and other resistances. Since no energy is supplied in free vibration, the motion in free vibration will diminish with time, and is said to be *damped*."

It follows from these descriptions that a *damper* is an element which can be added to a system to provide forces which are resistive to motion, thus providing a means of energy dissipation. Assuming that this working definition will suffice for general use, the next area of interest is to generally describe the functional output of a damper. As with the definition of damping, the functional output of a damper is somewhat controversial, since different output equations exist within the context of the various engineering disciplines.

The most convenient and common functional output equation for a damper comes from classical systems theory, and is that of the so-called *"linear"* or *"viscous"* damping element:

$$F = C\dot{X}$$

Where F = resistive force from the damping element
C = the damping constant
$\dot{X}$ = end to end velocity across the element

It is rather unfortunate that the engineers who established systems theory probably began first with electrical systems, where the functional output of a resistor follows a simple and linear form. Using a damper as the mechanical analog to a resistor caused dampers to be described in the same, linear manner. In the days before computers, system problems had to be solved simply, due to the slow calculating speed of the slide rule. With a linear expression for damping, many differential equations could be solved by manipulation and cancellation of terms, allowing for an economic solution time.

Conversely, in mechanical engineering, it is difficult to manufacture a useable fluid filled component having a purely viscous output, because even moderate pressure hydraulic flows through a simple orifice follow a much different output, in which differential pressure varies with the fluid velocity squared. The resultant hydraulic component varies its output force with respect to the squares law, and is the so-called *hydraulic damping element*, or *dashpot*. The output of the basic hydraulic damping element is:

$$F = C\dot{X}^2$$

Decades ago, when engineers first began contemplating and analyzing systems with self-contained fluid dampers, the first problem to be solved was to develop a damper design that would provide an output more in keeping with systems theory than with the realities of simple hydraulic flows. Evidently the funding was going to the analysis team first, then trickling down to the design team. Because of the large amounts of research funding required, high performance damping products were eventually developed with defense budget funds, with public disclosure and use of the technology being restricted. The results of this research were several damper internal constructions that made it possible to achieve, or at least mimic, the desired linear output from systems theory. Eventually both analytical methods and damper designs evolved to provide more optimal solutions to shock and vibration problems, with dramatic advances occurring in the 1960's and 1970's. As a result, the dampers being used today in buildings are of the so-called "low exponent" type, with an output equation of the form:

$$F = C\dot{X}^\alpha$$

In most cases, α is an exponent having a specified value in the range of 0.3 to 1.0. Values of α, which have proven to be most popular, are in the range of 0.4 to 0.5 for present-day building designs with seismic inputs. Bridge applications in U.S. Seismic Zones 3 and 4 use similar damping exponent values. Bridge applications in lessor seismic zones have often utilized an exponent of 2, that of the classical hydraulic damping element. Wind damping applications presently are most popular with exponents in the range of 0.5 to 1.0, with the lower values being used in structures driven by both wind and seismic inputs.

Other types of dampers exist which have very different outputs from viscous or fluid damping devices. These include so-called friction or hysteretic dampers, and rubber or visco-elastic dampers. A friction damper is essentially an on-off constant force device, where the resistive force to any motion, large or small, is a single fixed value. Rubber damping elements are relatively complex, and indeed no single output function exists to define the performance of rubber damping elements. The actual output varies with the type of rubber, how the rubber is shaped and constrained, and ambient temperature. In general, a rubber damping device can be modeled as a spring element in a series with a Voight element; the Voight element consisting of a spring and damper element in parallel with each other. In most cases, both hysteretic and rubber damping devices are only used where relatively small amounts of damping are required in a structure, usually less than 5% critical. The reasons for this will be discussed later on in this section.

Generalized Effects Of Adding Damping To A Structure

Damping is one of many different methods that have been proposed for allowing a structure to achieve optimal performance when it is subjected to seismic, wind storm or other types of transient shock and vibration disturbances. Conventional approach would dictate that the structure must passively attenuate or dissipate the effects of transient inputs through a combination of strength, flexibility, deformability and energy absorption. The level of damping in a conventional structure is very low, and hence the amount of energy dissipated during transient disturbances is also very low. During strong motions, such as earthquakes, conventional structures usually deform well beyond their elastic limits, and remain intact only due to their ability to inelastically deform. Therefore, most of the energy dissipated is absorbed by the structure itself through localized damage.

The concept of added-on dampers within a structure assumes that some of the energy input to the structure from a transient will be absorbed, not by the structure itself, but rather by supplemental damping elements. An idealized supplemental damper would be of a form such that the force being produced by the damper is of such a magnitude and occurs at such a time that the damper forces do not increase overall stress in the structure. Properly implemented, an ideal damper should be able to simultaneously reduce both stress and deflection in the structure.

Figure 1 depicts earthquake spectra capacity and demand curves for a sample building with 20%, 30% and 40% damped demand curves. This Figure is reproduced from FEMA 274 [2] and assumes linear or viscous damping elements are used.

The effects of added supplemental damping in a structure subjected to earthquake transients is depicted in the test results provided in Figures 2 and 3. The tested structure was a single story, steel building frame, using steel moment frame connections. Figure 2 shows the response of the test structure under a scaled input of 33% of the 1940 El Centro earthquake. Note that a small hysteresis loop is apparent in Figure 2, revealing that the test structure was at the onset of yield. Structural damping in the frame was in the 1½ - 2% range. In comparison, Figure 3 is the same structure with 20% added damping, obtained by the addition of two small linear fluid dampers installed as diagonal brace elements. The large energy dissipation of added damping is readily apparent in the football shaped damping curve superimposed over the structural spring rate curve. Note also that the input in Figure 3 is the full 100% El Centro earthquake, yet base shear and deflection of the frame are virtually unchanged from the undamped case of Figure 2. Thus, in this case, the addition of 20% added linear damping to the structure increased its earthquake resistance by a factor of 3, compared to that of the same structure without added damping. Most importantly, this threefold performance improvement was obtained without increasing the stress or deflection in the structure. In fact, it is this tremendous performance improvement that has caused much of the interest in fluid dampers for structural engineering use. To paraphrase the body builder's saying, this is a case where dampers provide a big gain, without any pain!

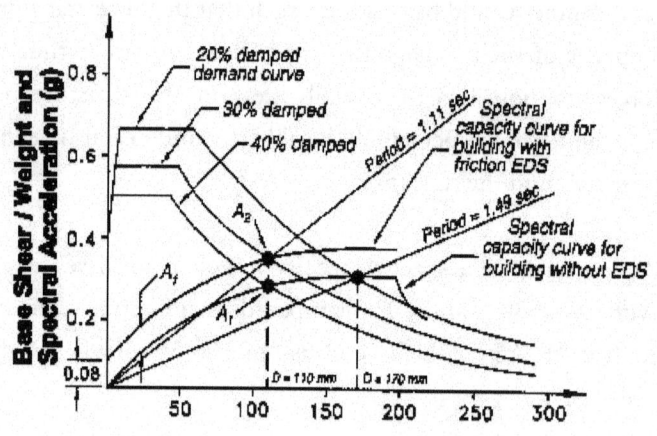

Lateral Deformation and Spectral Displacement (mm)

FIGURE 1

SPECTRAL CAPACITY AND DEMAND CURVES FOR REHABILITATED ONE-STORY BUILDING

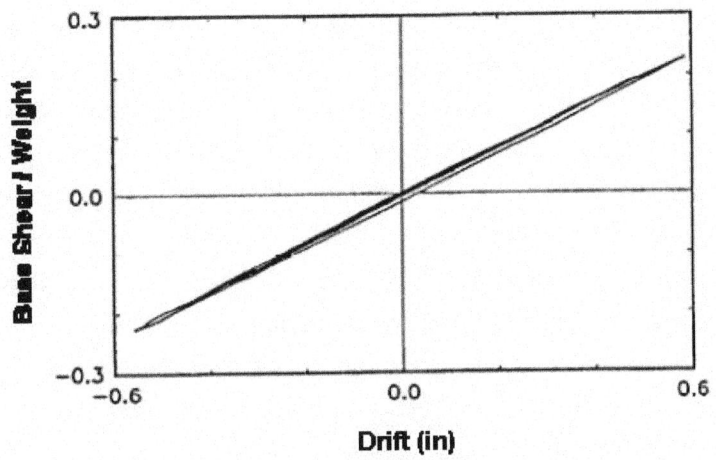

FIGURE 2

ONE-STORY STRUCTURE, NO DAMPERS B EL CENTRO 33.3%

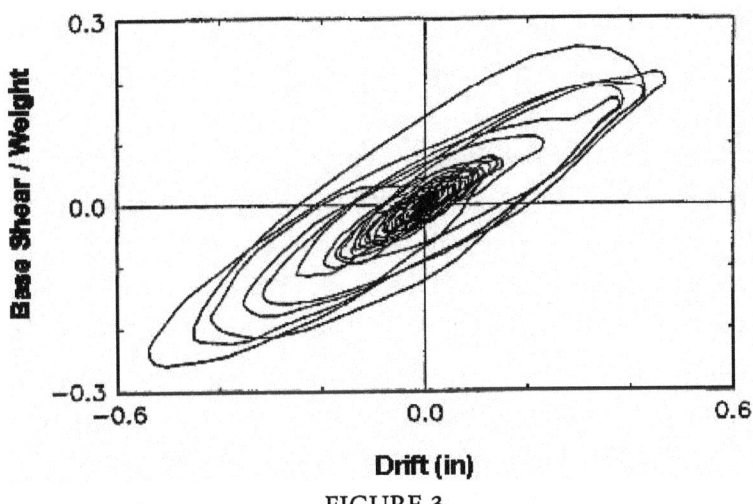

FIGURE 3

ONE-STORY STRUCTURE, TWO DAMPERS B EL CENTRO 100%

The test results from Figures 2 and 3 used the 1940 El Centro earthquake transient as a test input. When these results were first published in 1992 by

Constantinou and Symans [3], they included tests showing similar performance gains with other notable earthquakes for which transient records were available. Nevertheless, questions have arisen in the ensuing years as to whether fluid dampers would be functional with other inputs, including actual earthquakes such as the 1994 Northridge, California and 1995 Kobe, Japan events, plus hypothetical inputs such as "a big, purely impulsive quake" or "a slow rolling sine wave quake." In addition, potential customers with wind storm inputs wanted to know if seismic dampers worked in wind, and Government customers wanted to know if damage from terrorist attacks against buildings would be reduced by dampers. The actual question being raised was simply: "Fluid dampers appear to be a useful engineering component. Are they truly useful for all types of shock and vibration inputs?" The answer is a definite yes, and it is relatively easy to demonstrate this by considering generalized qualities of a transient pulse.

The first and most important parameter of a transient is the peak translational velocity. The peak velocity is of primary importance because this determines the peak amount of energy that must be managed by the structural system. This velocity can be achieved by either a small acceleration over a long time period, or by a large acceleration over a short period. Thus, the maximum acceleration rate of the pulse is the second most important parameter of a transient, since the structure and the fluid dampers must be designed to accommodate the acceleration without being damaged by impulsive loadings. Figure 4 provides tabular data for maximum velocities and accelerations for catastrophic inputs. The least important parameters of the transient are those related to the actual shape of the various portions of the pulse. This is simply because no two discrete transients can be expected to be identical, these events being chaotic by their very nature. If one considers how a damped structure behaves under transients having a given maximum translational velocity and maximum acceleration then, in reality, only two simple extreme cases need to be considered.

Case One: The structure is excited by a step function, with acceleration equal to the maximum acceleration expected, for a time duration such that maximum translational velocity is obtained.

DAMPERS: HOW VISCOUS DAMPERS PROTECT STRUCTURES

Case Two: The structure is excited by a forced sine wave at the frequency of the structure=s first resonant mode, with input amplitude increased until the maximum specified acceleration or velocity is achieved.

An example of structural response to the first case, the impulsive input, is provided in Figure 5, for both the undamped and fluid damped condition. The response in this case assumed infinite acceleration, with velocity stepping from zero to maximum value instantaneously, and an elastic structure. It is readily apparent that the fluid damped structure experiences substantially less force and deflection than the undamped structure, even though each structure is storing or absorbing equal amounts of impulse energy.

An example of the second case is provided in Figure 6, from Thomson [1], and depicts the magnification factor on input amplitude for a system subjected to forced harmonic excitations with linear fluid damping. The condition of resonance is obtained at a frequency ratio of 1.0, and shows the tremendous benefits of fluid damping. The equation for magnification at resonance is:

$$\text{magnification factor} = \frac{1}{2\zeta}$$

where ζ = the damping ratio

Of particular note is that for a typical building with 2% damping, the magnification factor at resonance is 25 to 1. This number reduces to a much more manageable value of only 2 to 1 at 25% damping. It is of value to the engineer to note that virtually no structure is built with the safety factor of 25 to 1 necessary to accommodate the 2% damped resonant response. In comparison, most structures have sufficient safety factors to accept the 2 to 1 magnification for the 25% damped structure subjected to forced resonance.

From these examples, it is relatively easy to understand that fluid damping will always improve the response of a structure, under any expected transient.

TABULAR DATA FOR MAXIMUM VELOCITIES AND ACCELERATIONS		
	Peak Acceleration	Peak Velocity
Northridge Earthquake	.9 G	51 In/Sec.
Kobe Earthquake	.8 G	35 In/Sec.
Ship, Moored Mine	25. G+	90 In/Sec.+
Missile Silo, Nuclear Air Burst	80. G+	450 In/Sec.+
Submarine, Nuclear Depth Charge	600. G+	500 In/Sec.+

FIGURE 4

CATASTROPHIC TRANSIENTS

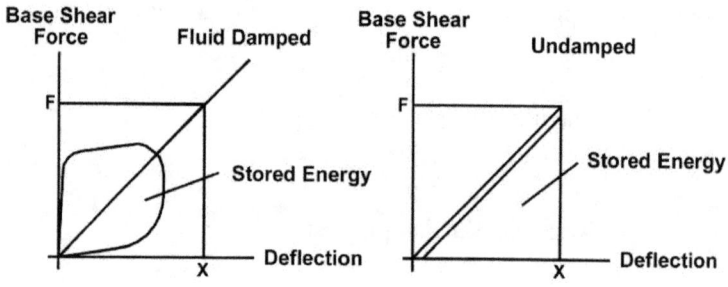

FIGURE 5

RESPONSE TO IMPULSIVE INPUTS

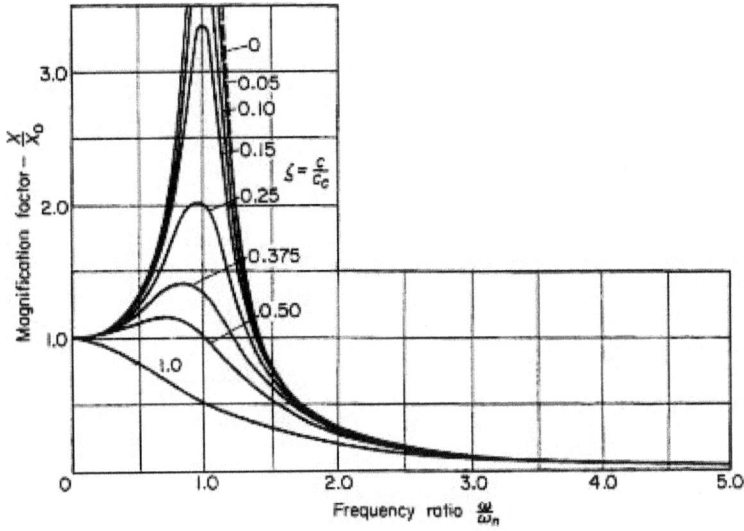

FIGURE 6

MAGNIFICATION FACTOR FOR FORCED HARMONIC EXCITATION

Three Generic Types of Dampers and How Each of Them Affects a Structure

Fluid dampers have the unique ability to simultaneously reduce both stress and deflection within a structure subjected to a transient. This is because a fluid damper varies its force *only* with velocity, which provides a response that is inherently out-of-phase with stresses due to flexing of the structure. Other dampers can normally be classified as either hysteretic, where a fixed damping force is generated under any deflection, or as visco-elastic, where a damper behaves as a complex spring and damper combination. In the latter case, force may be a displacement *and* velocity dependent parameter. Figure 7 provides representative outputs from sine wave excitation of these three damper types. Inclusive in these non-fluid damper types are yielding elements, friction devices, plastic hinges, friction slides, bonded rubber, molded rubber, and shaped rubber. None of these other devices have an out-of-phase response to structural flexural stresses. This is simply because the outputs of these devices are dependent upon parameters other than, or in addition to, velocity. Hence, all of these other types of dampers will decrease deflection in a structure at the same time they are increasing column stress. The out-of-phase response that is unique to fluid dampers can be easily understood by considering a building shaking laterally back and forth during a seismic event or a windstorm.

Column stress is at a peak when the building has flexed a maximum amount from its normal position. This is also the point at which the flexed columns reverse direction to move back in the opposite direction. If we add a fluid damper to the building, damping force will reduce to zero at this point of maximum deflection. This is because the damper stroking velocity goes to zero as the columns reverse direction. As the building flexes back in the opposite direction, maximum damper force occurs at maximum velocity, which occurs when the column flexes through its normal, upright position. This is also the point where column stresses are at a minimum. It is this out-of-phase response that is the most desirable design aspect of fluid viscous damping.

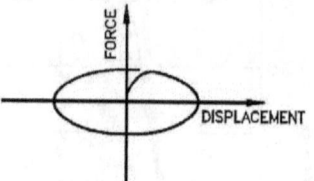

Fluid Viscous Output

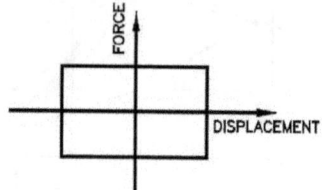

Friction Output

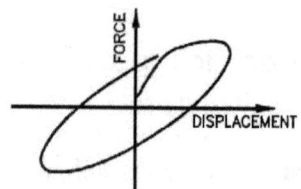

Viscoelastic Output

FIGURE 7
OUTPUT OF THE THREE
GENERIC DAMPER TYPES
FLUID DAMPING DEVICES: A
CENTURY OF HISTORY

It is axiomatic that during times of war, new technology develops extremely quickly, since the fates of nations may well depend upon which antagonist can mass-produce improved weapons more quickly. In the case of fluid dampers, the evolution of large bore artillery and naval guns in the late 1800's provided the need for the product, and the various major governments were only too eager to provide the development funding.

The Guns of War, 1897-1918 - Necessity Fosters Invention

The evolution of large dampers began with the advent of large breech loaded cannons in the 1860's. Prior to this, large guns were muzzle loaded in a very time-consuming manner. Gaining easy access to the gun's muzzle end for loading was simple, the weapon was merely allowed to move backwards anywhere from one to twenty feet after firing. Motion was retarded by means of a spade-like device literally digging into the earth on land-based weapons. Shipboard guns used friction slides or inclined surfaces to arrest their firing motion, often aided with block and tackle mechanisms. After loading, the gun crew would push the gun back into its battery, or ready to fire position. The advent of breech loading allowed for much more rapid (and safer) loading of the weapon, and a desirable higher rate of fire. Unfortunately, the high firing rate required that the gun crew work much faster repositioning the gun, quickly exhausting the crew.

Several unsuccessful concepts of arresting gun recoil were attempted, involving both coil springs and rubber blocks. Meanwhile, the inventors of that time were investigating the new field of hydraulic components, and by the late 1860's, experiments were taking place using hydraulic dampers to arrest gun recoil. It is reported by Hogg [4] that the British Army was the first to use hydraulic recoil dampers on gun carriages in 1862. The first mass-produced hydraulic recoil damper was used on the 75 mm French field gun, Model M1897. This weapon was hailed as a true technological marvel and is considered to be the first modern artillery piece. The carriage of the weapon included a slide to support the gun itself, and a 48 inch stroke fluid damper combined with a light spring to attenuate recoil energy and return the gun to battery. The French M1897 went on to serve in both World War I and World

War II. Many variations of the weapon exist since many countries "borrowed" the design after capturing one or more examples during World War I. One of the more unusual uses for the low recoil French M1897 was by the U.S. Army Air Corps during World War II. The Air Corps needed a ground attack aircraft with as much firepower as possible. The solution to the problem involved mounting a complete M1897 with recoil dampers into the nose of the U.S. Model B-25 "Mitchell" Bomber, firing forward. The modified aircraft proved successful, and the use of the hydraulic dampers eliminated damage to the aircraft.

By the end of World War I, tens of thousands of fluid dampers were being used on field artillery pieces, naval guns, coastal guns and railway guns. Some dampers of this period were even of the semi-active type, where changing the gun elevation angle would change the resultant damping force. This was accomplished by using a gear train between gun carriage and the damper. The gear train would rotate an adjustment rod or screw protruding from the damper cylinder. As the gun was elevated, the damper would become stiffer, and use less displacement. This feature allowed the gun carriage to be reduced in size and weight, since at high elevation angles, the carriage no longer needed to maintain clearance to the ground for the entire recoil stroke.

Toward the end of World War I, another advantage of fluid dampers was discovered. This was that reduced recoil allowed weapons to easily fire larger projectiles, with larger propellant charges to obtain greater range. Indeed, from March to July of 1918, the City of Paris was attacked by the German Army with a weapon of "super gun" proportions. Details did not become available until the war ended, and then only after intense efforts by the allies. The weapon was named the Paris Gun, and included a 130 foot long barrel, which fired a 210 mm diameter shell at a range up to 85 miles. The gun itself, with fluid dampers, weighed over 140 tons, not including the weight of the tremendous carriage that carried the weapon. Three of the Paris Guns were built, but all were withdrawn from service as the allied armies approached their locations. Mysteriously, none were recovered by the allied forces after the war ended.

The Automotive Damper - Optimization Through Evolution

The 1920's and 1930's were a period when the automobile became a dominant feature of American culture. Since the automobile was a relatively new product with a large potential market, automotive manufacturers were forced, by competitive pressures, to produce a product that would be appealing to the consumer. One of the most appealing traits that an automobile could possess was a smooth ride over all possible road surfaces; this proved to be a true challenge for automotive engineers of this period.

The earliest auto suspensions were simply carried over from horse-drawn wagons. The suspension consisted of multiple leaf elliptical or semi-elliptical springs. Damping was limited to the inter-leaf friction which occurred as the spring leaves ground over one another as the spring deflected. Damping would obviously have a high variance from day-to-day, depending on whether the spring was dry, wet, rusty, dirty, or recently cleaned and oiled.

This day-to-day damping change proved unacceptable to the consumer, and external friction pad or rubber dampers were added to the suspension. These provided a small but noticeable improvement over using the spring itself as a damper, plus it was possible to make the damper adjustable for wear. The "ideal" damping material was usually pure asbestos washers or pads, compressed between two iron plates. One plate was fixed to the car frame by a bolt, the other was attached to an actuating arm. A large draw bolt went through the center of the damper assembly, and tightening or loosening of the bolt served to adjust the damping force.

The high maintenance and marginal improvement obtained with friction and rubber dampers caused automotive parts suppliers to look for improved damping systems, and fluid dampers quickly entered the scene. The biggest problem with adapting the fluid damper for automotive use proved to be poor quality seals. The guns of World War I usually needed a major overhaul every 500 rounds or so, due to barrel wear, and this was an opportune time to change damper seals, which usually were leaking badly after 500 cycles.

Considering that the seals of the day consisted of cut lengths of hemp rope forced into a pocket with a hammer, this was no surprise! "Improved" seals of the 1920's consisted of a stack of round leather washers forced into position with a packing nut. These were an improvement over hemp strands, but still could not provide the cyclic life necessary for automotive use.

In 1925, Ralph Peo of the Houdaille Company in Buffalo, New York, invented a solution to the seal problem. Instead of improving the seal, he redesigned the damper to use a rotating piston rod and vane assembly, thus replacing long travel, sliding seal motion with a short 60-120 degree rotary travel. The Houdaille rotary damper was actuated by crank arms attached to the moving components of the suspension. The short rotary travel of the seal allowed for roughly 10,000 miles of road travel before seal replacement was necessary. Within a short period, most automobiles were using the Houdaille rotary damper. Figure 8 is one of the original patent sheets depicting Peo's 1925 invention.

In 1949, the Delco Division of General Motors finally designed a sliding seal damper that had an adequate life for automotive use, thus ending the rotary damper era. Present-day automotive shock absorbers have an internal construction that is very similar to the gun recoil buffers of World War I, except that modern seals provide substantially greater life.

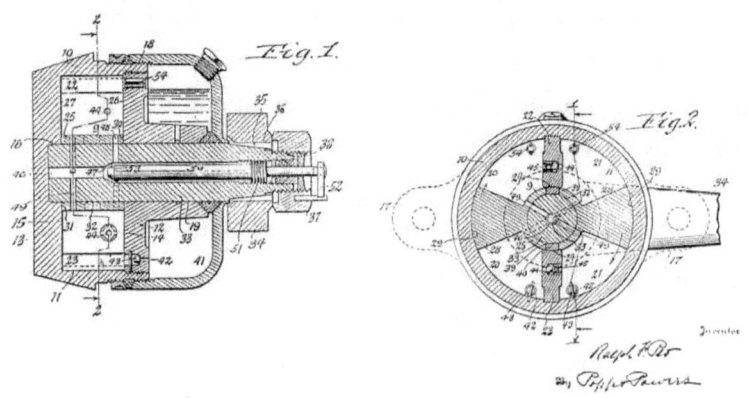

FIGURE 8
PATENT SHEET – R. PEO'S ROTARY SHOCK ABSORBER

The Cold War - Dampers Go Underground

History texts will eventually include great amounts of information about the Cold War period, which lasted from the end of World War II to approximately 1990. Detailed information will be presented only as it is declassified, and this will take many years. Early on in the Cold War, both the United States and Russia began developing intercontinental ballistic missiles (ICBM), equipped with nuclear warheads. Although still debated, most defense analysts state that the U.S. strategic war doctrine was such that our missiles would not be launched until enemy warheads had actually detonated on or above U.S. soil. Adherence to this doctrine assumed that the enemy=s initial targets would be U.S. missile launchers, striking as many of these as possible in a first strike. In order for the U.S. to launch a counterstrike under these conditions, our missiles needed to be designed and/or based in such a way that they could survive a nuclear attack without damage. Initially, land based missiles were simply placed underground in heavily reinforced launch silos, usually accompanied by underground launch facility buildings. However, as missile guidance systems evolved, the accuracy of enemy missiles was improved, and the need for shock isolation devices became apparent. Early missile isolators consisted of simple coil springs with fluid dampers. In some cases, the spring-damper units were used to isolate the missiles themselves and various critical items inside the launch complex. In other cases, entire structures were base isolated in vertical and horizontal planes.

During the 1960's, it became impossible to provide large enough mechanical springs to provide the optimal isolation, so fluid dampers were converted to liquid-spring dampers, an extremely powerful yet compact isolation component. In a liquid spring-damper, the operating fluid is compressed and orificed simultaneously. By selecting special fluids with high compressibility, it was possible to produce both high spring and damping forces in an extremely small package. Without becoming too specific (for security considerations), some of the liquid spring-dampers of the late 1980's could simultaneously provide spring forces of 50 tons and damping forces of 150 tons from a package of only seven inches in diameter! Operating fluid pressures of up to 50,000 lb/in² were relatively common. In comparison, a high-powered hunting rifle has peak firing pressures in the 40,000 lb/in² range. Some of these products for large land based missiles had more than four feet of displacement, with output forces up to 500 tons.

The successful use of high capacity fluid dampers and liquid spring-dampers on land based missile facilities led to additional applications on shipboard and submarine missiles and related equipment items. By the end of the Cold War, a typical U.S. Naval warship would have more than 1,000 fluid damping devices installed on its missiles and primary electronics systems. These devices range from 1 ton to 50 tons of output force.

During the 1990's, the end of the Cold War combined with the political and economic climate caused a dramatic downsizing of U.S. defense capabilities. At the same time, security restrictions on the sale and commercial use of Cold War era technology had been greatly relaxed.

The Last 30 Years - Transition of Defense Technology to the Private Sector

Defense firms found very few new opportunities in their traditional markets when the Cold War ended. Some firms grew smaller, or maintained sales levels by oftentimes painful mergers or consolidations. Relatively few firms were able to transition their technology to the commercial marketplace. Taylor Devices, Inc., a New York based manufacturer of energy absorption products for military and defense use, began to look for commercial outlets for its defense products in 1987.

Taylor Devices= defense expertise involved the design and manufacture of large, fluid damping devices for protection of missiles, electronics systems, and large structures against the effects of weapons explosion. The company=s staff elected to pursue commercial applications related to seismic and high wind protection of structures. The damper style selected dated from the

1970's, and was developed on a sole-source basis by the firm for use on the U.S. Air Force's MX Ballistic Missile, and the U.S. Navy=s Tomahawk Cruise Missile. On the latter program, the company has produced more than 29,000 fluid damping devices for use on the shipboard launched Tomahawk.

Early on, it was decided to pursue joint research on fluid damped building and bridge structures with the National Center for Earthquake Engineering Research (NCEER). NCEER was conveniently located on the campus of the State University of New York at Buffalo, just a short distance from Taylor Devices' facilities. The research involved taking existing military production fluid damping devices, and simply installing them onto scaled models of civil engineering structures, as supplemental components. The structures were then subjected to seismic transient testing on the University's large seismic shake table. All tests proved excellent, with dramatic reductions of stress and deflection occurring with added fluid damping in the 15-40% range.

In general, it was found that adding 20% damping to a structure will triple its earthquake resistance, without increasing stress or deflection. Numerous reports were published by NCEER and the University, documenting the improvements obtainable with fluid dampers. The U.S. Department of Defense proved very cooperative in allowing Taylor Devices to disclose the origins and applicable design concepts for the damping devices used in the research.

For example, steel building structures were tested with fluid dampers being currently produced for the B-2 Stealth Bomber. Concrete building structures were tested using Tomahawk missile dampers. Bridge structures were tested with dampers from the CIA's famed Glomar Explorer Research Vessel. Other bridge structures were fitted with spring-damper units from submarine based torpedoes.

It became evident that there were no barriers towards commercial implementation of Taylor's damping products, and by 1993, an order was received for 186 dampers to be used on all five buildings of the new Arrowhead Regional Medical Center in Colton, California. Specifications for these dampers are provided in Figure 9, and a photo of a completed damper follows in Figure 10.

No design or development was necessary by Taylor Devices to build these large dampers, even though each device produces 160 tons of force and has a 48 inch displacement. The reason was simply that it already was a production design, used as the vertical shock isolator for the U.S. Air Force MX Ballistic Missile, installed in the 1978 Multiple Protective Shelter basing mode dating to the Carter Presidential Administration.

More than 700 additional building and bridge projects followed the Arrowhead Medical Center order. The transition of fluid dampers from military to civilian has proven to be the quintessential example of literally "turning swords into plowshares."

Displacement	= 48 in.
Maximum Damping Force	= 320,000 lb.
Maximum Operating Velocity	= 60 in/sec.
Power Dissipation	= 2,170,000 watts
Length	= 14.5 ft. extended
Diameter	= 14 in.
Weight	= 3,000 lb.
Quantity Required	= 186 units

FIGURE 9

SAN BERNARDINO COUNTY MEDICAL CENTER DAMPER SPECIFICATIONS

FIGURE 10
PHOTOGRAPH OF COMPLETED DAMPER

FLUID DAMPERS FOR BUILDING AND BRIDGE STRUCTURES

The essential design elements of a fluid damper are relatively few. However, the detailing of these elements varies greatly and can, in some cases, become both difficult and complex. Figure 11 depicts a typical fluid damper and its parts. It can be seen that by simply moving the piston rod back and forth, fluid is orificed through the piston head orifices, generating damping force.

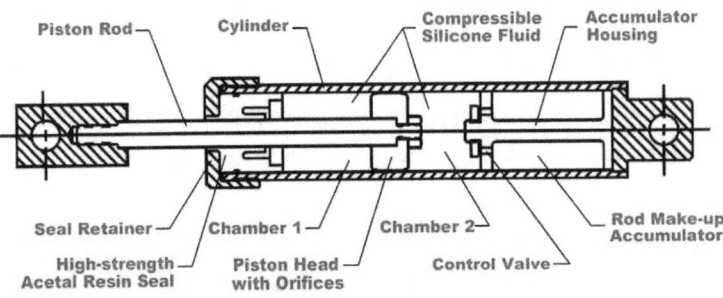

FIGURE 11 FLUID DAMPER

Major part descriptions are as follows, using Figure 11 as reference:

Piston Rod Highly polished on its outside diameter, the piston rod slides through the seal and seal retainer. The external end of the piston rod is affixed to one of the two mounting clevises. The internal end of the piston rod attaches to the piston head. In general, the piston rod must react all damping forces, plus provide a sealing interface with the seal. Since the piston rod is relatively slender and must support column loading conditions, it is normally manufactured from high-strength steel material. Stainless steel is preferred as a piston rod material, since any type of rust or corrosion on the rod surface can cause catastrophic seal failure. In some cases, the stainless steel must be chrome plated for compatibility with the seal material. In addition, the design of the piston rod should be strain based, rather than stress based, since elastic flexing of the piston rod during damper compression can cause binding or seal leakage. Bending loads on the piston rod can become a design issue if a damper has more than 12 inches of displacement. For applications requiring a long stroke, a structural steel tube guide sleeve is used to protect the piston rod from excessive bending loads. The Arrowhead Medical Center damper, shown previously in Figure 10, incorporates a guide sleeve of this type.

Cylinder The damper cylinder contains the fluid medium and must accept pressure vessel loading when the damper is operating. Cylinders are usually manufactured from seamless steel tubing. Welded or cast construction is not permissible for damper cylinders, due to concerns about fatigue life and stress cracking.

Cylinders normally are designed for a minimum proof pressure loading equal to 1.5 times the internal pressure expected under a maximum credible seismic event. By definition, the proof pressure loading must be accommodated by the cylinder without yielding, damage, or leakage of any type.

Fluid Dampers used in structural engineering applications require a fluid that is fire-resistant, non-toxic, thermally stable, and which will not degrade with age. Using current Occupational Safety and Health Administration (OSHA) guidelines, this fluid is classified as non-flammable and non-combustible to OSHA Class IIIB, with a fluid flashpoint above 200 degrees F. At present, the only fluids possessing all of these attributes are from the silicone family. Typical silicone fluids have a flashpoint in excess of 650 degrees F, are cosmetically inert, completely non-toxic, and are among the most thermally stable fluids known to man. Since silicone fluids are produced by distillation, the fluid is completely uniform and no long-term settling will occur. The typical silicone fluid used in a damper is virtually identical to the silicone used in common hand and facial cream cosmetics.

Seal The seals used in a fluid damper must be capable of a long service life; at least 25 years without requiring periodic replacement. The seal materials must be carefully chosen for this service life requirement and for compatibility with the damper's fluid. Since dampers in structures are often subject to long periods of infrequent use, seals must not exhibit long-term sticking nor allow slow seepage of fluid. Most dampers use dynamic seals at the piston rod interface, and static seals where the end caps or seal retainers are attached to the cylinder. For static seals, conventional elastomer o-ring seals have proven to be acceptable. Dynamic seals for the piston rod should be manufactured from high-strength structural polymers, to eliminate sticking or compression set during long periods of inactivity. Typical dynamic seal materials include Teflon, stabilized nylon, and members of the acetyl resin family. Dynamic seals manufactured from structural polymers do not age, degrade, or cold flow over time. In comparison, conventional elastomers will require periodic replacement if used as dynamic seals in a damper.

Piston Head The piston head attaches to the piston rod, and effectively divides the cylinder into two pressure chambers. As such, the piston head serves to sweep fluid through orifices located inside it, thus generating damping pressure. The piston head is usually very close fit to the cylinder bore; in some cases the piston head may even incorporate a seal to the cylinder bore.

Seal Retainer Used to close open ends of the cylinder, these are often referred to as end caps, end plates, or stuffing boxes. It is preferable to use large diameter threads turned on either the exterior or interior surface of the cylinder to engage the seal retainer. Alternate attachment means, such as multiple bolts, studs, or cylinder tie rods should be avoided as these can be excited to resonance by high frequency portions of either the earthquake transient or the building response spectra. Tie rods should not be used since they generally constitute an unacceptable single point failure; a catastrophic design flaw. If even a single tie rod yields during a seismic event, the seal retainer will usually bend or rotate such that an open gap appears between the cylinder and retainer. The damping fluid is literally pumped out of the damper through this open gap. Thus, the mere yielding of a single tie rod can cause a total loss of all damping output. This problem does not occur with a seal retainer that is threaded directly into the cylinder bore. In this case, if a single thread yields, the pressure loading is distributed among the numerous remaining threads.

Figure 11 depicts a damper with a seal retainer threaded to the exterior of the cylinder. Figure 12 depicts a cylinder with an internally threaded retainer. Figure 13 depicts an unacceptable construction with external tie rods. Observation of Figures 12 and 13 easily reveals that the tie rods constitute a weak point in the design, with relatively small cross section and low elasticity.

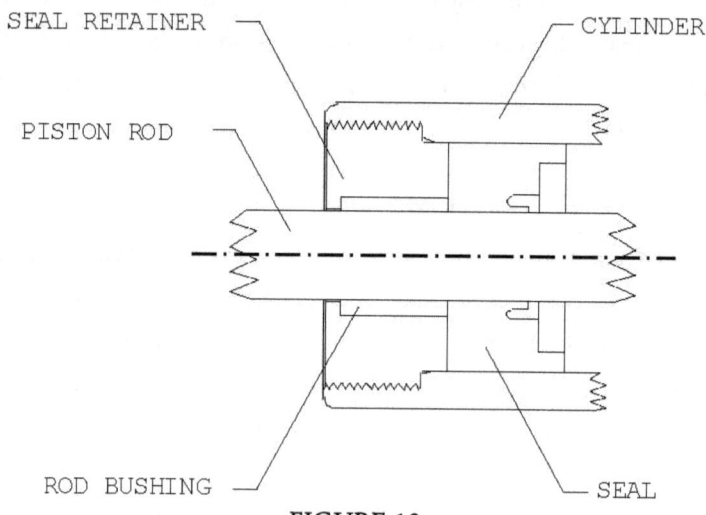

FIGURE 12
INTERNALLY THREADED CYLINDER WITH RETAINER

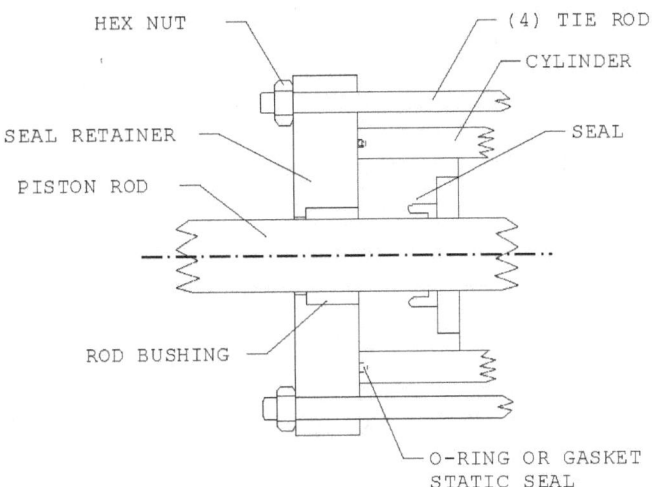

FIGURE 13

TIE ROD CYLINDER WITH RETAINER

Accumulator The simple damper depicted in Figure 11 utilizes an internal in-line rod make-up accumulator. The accumulator consists of either a block of closed cell plastic foam, a moveable (and gas pressurized) accumulator piston, or a rubber bladder. The purpose of the accumulator is to allow for the volumetric displacement of the piston rod as it enters or exits the damper during excitation. A second purpose is to compensate for thermal expansion and contraction of the fluid. The damper in Figure 11 uses a control valve to meter the amount of fluid displaced into the accumulator when the damper is being compressed. When the damper extends, the control valve opens to allow fluid from the accumulator to freely enter the damper pressure chambers. Although the damper in Figure 11 has an in-line, internal accumulator, some older damper types use an external accumulator tank with connecting hoses or piping.

Orifices The pressurized flow of the fluid across the piston head is controlled by orifices. These can consist of a complex modular machined passageways, or alternately, can use drilled holes, spring loaded balls, poppets, or spools. Relatively complex orifices are needed if the damper is to produce output with a damping exponent of less than two. Indeed, a simple drilled hole orifice will follow Bernoulli's equation, and damper output will be limited to varying force with the square of the damper velocity. Since "velocity squared" damping is of limited use in seismic energy dissipation, more robust and sophisticated orifice methods are usually required. One method uses a patented series of precisely shaped orifice passages, making use of the scientific processes of fluidic controls. Dependent on the shape and area of these passages, damping exponents ranging from 0.3 to 1.0 can be obtained without requiring any moving parts in the orifice.

The use of any type of spring loaded orifice mandates that rigorous full-scale testing be performed on the damper at the maximum expected velocity and frequency for the application. Scale models or generic testing cannot be used for performance validation. The reason for this is simply that it is not possible to scale hydraulic relief valves. This scaling problem exists because both the orifice flow through the valve and the valve spring forces will vary with the *square* of their respective diameters. At the same time, the weight of the ball, poppet, or spool element used as the valve closure element varies roughly with the *cube* of the diameter. Therefore, as an orifice is made larger, its performance under impulsive inputs and its potential frequency response range is degraded, due to the valve closure element becoming proportionately heavier relative to the rest of the valve. Thus, to maintain proper performance requires that each damper have its own dedicated valve design, which does not scale from other sizes. This, combined with reliability issues surrounding multiple spring loaded valves, mandates full scale testing of each and every damper produced.

Figure 14 is a test curve from a relief valve type orifice used in a 125 kip output force damper. The relief valve is sticking in its bore in the extension direction. Following the curve from left to right shows the compression side valve opening properly, and maintaining a force level of 125,000 lbs. plus or minus 15%, per specification. When the test machine begins to pull the damper in extension, the anomaly occurs. When the valve begins to stroke, it sticks in its bore and "overshoots" the nominal 125 kip output to a value of 145 kip. When the valve fully opens, it continues to bind in its bore, with the force dropping off to 100 kip. Disassembly of the valve found a small machining burr on the valve spool that was causing the binding condition. The burr was removed, the damper was reassembled, and tested acceptably. If the damper in question had not been tested individually, an unacceptable defective product would have been delivered to the job site.

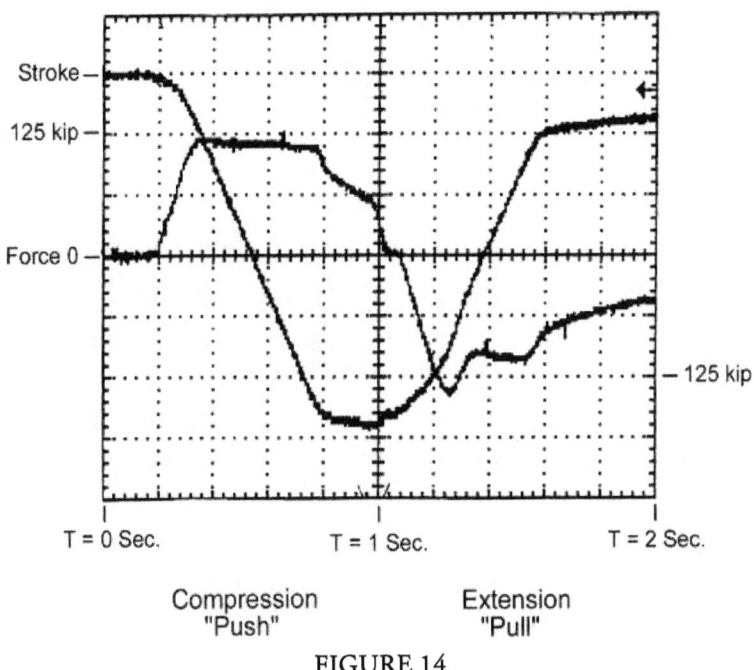

FIGURE 14

TESTING OF DAMPER WITH DEFECTIVE RELIEF VALVE

Operational Methods

The basic damper design of Figure 11 is essentially generic, in that it depicts all of the basic design elements required for proper component output. Specific types of dampers may rearrange or alter some of the basic elements depicted, based upon customer or manufacturer=s preferences.

One alternate design of damper utilizes a piston rod that goes entirely through the cylinder of the damper, so that there is no net volumetric displacement of the piston rod as the damper strokes. In this type of design, the accumulator becomes much smaller, since it is only needed for thermal compensation of the fluid volume. This particular damper type can also eliminate the accumulator control valve, and often uses only a small metering hole into the accumulator to allow small thermal expansion/contraction volumes of fluid to enter/exit the accumulator. This style of damper is known as a "through-rod" or "balanced rod" damper.

Operating Pressures

The operating pressure of a damper influences its physical envelope and the material properties of its internal parts. These, in turn, influence the cost of the device. Currently available seismic dampers appear to be most cost-effective when designed to operate at 5,000 - 8,000 psi maximum pressure. In comparison, wind damper designs must be capable of continuous energy dissipation during storms lasting several hours. In most cases, this means that a wind damper cannot operate above 2,000 psi, or overheating of the damping fluid and dynamic seals can occur.

In comparison to current military and aerospace fluid damping devices, seismic and wind dampers use extremely conservative internal operating pressures. When cost is not of particular concern (as compared to minimum size and low weight), internal operating pressures of non-commercial, high performance dampers can range from 30,000 to nearly 50,000 psi. These military and aerospace products are similar in design and appearance to commercial dampers, but typically use much more costly construction materials. Some of these often include maraging or precipitation hardening steels, having yield strengths in excess of 200,000 psi.

Materials of Construction

Fluid Dampers are essentially fluid filled mechanisms which must be capable of extremely long term service without maintenance. In addition to requiring materials that are inherently corrosion resistant, damper materials have additional constraints, including low notch sensitivity, freedom from stress cracking, and a high impact resistance. This is especially true for the cylinder of the damper, which is a pressure vessel, and must accept substantial tri-axial stresses.

Within American industry, various standards for materials exist, maintained by various independent organizations. Some of these, such as the American Society for the Testing of Materials (ASTM), merely list the performance specifications for a specific material, making no comment as to its capability or suitability for a specific end use. Another organization of this type is the American Iron and Steel Institute (AISI); responsible for specifying steel alloy designations, nomenclature, and chemical properties. Other standards organizations maintain lists of materials that have been tested in specific applications and have been found to be acceptable. Some of these organizations are listed below, with their applicable specifications:

1. Society of Automotive Engineers Aerospace Materials Specifications (AMS)
2. American Society of Mechanical Engineers AASME Standards
3. United States Department of Defense, MIL-Handbook 5, Metallic Materials and Elements for Aerospace Vehicle Structures
4. NASA, Goddard Space Flight Center Materials Selection Guide

It is easy to cite examples of why end use is of critical importance in the selection of any material for an engineered component. Consider the alloy steel designated as AISI Type 4140, listed in ASTM A29 for bar stock. The ASTM specification spells out the chemical composition allowed for the material, and tells the engineer how to chemically sample and tensile test the material. Hopefully, the ASTM limits are identical with the AISI requirements for that specific alloy, since the AISI is the cognizant organization for alloy designations. This particular material can be heat treated, by quenching and tempering, to yield strength levels as high as 250 ksi. This normally would imply that this material would be a good choice for a damper piston rod, with chrome plating added for corrosion protection and wear resistance. However, some of the other organizations noted above have some very negative comments on the use of this material in specific applications.

For example, NASA Goddard's Material Selection Guide states that this particular material in the strength level noted has a "low resistance to stress corrosion cracking," and that it "is not approved for any long term use involving high stress." The Department of Defense notes that this steel "may be embrittled by tempering or exposure to the 500 degrees F to 700 degrees F range," thus limiting its application to heat treatments providing less than 180 ksi yield. The Defense Department has additional comments concerning the use of this material with chrome plating. First, the material is not approved unless a specific time sensitive additional heat treat is performed after plating, and even with the extra heat treat, the material is not approved to carry any dynamic or non-static load. To carry dynamic loads, such as would be expected in use as a damper piston rod, the material must first be machined to specific limits, controlling the size of notches and radiuses. Then the material must be heat treated, then shot peened per a prescribed plan prior to plating, then heat treated a second time after plating.

Thus, in this case, a material listed in accordance with the appropriate ASTM standards appears to be unsuitable, and even potentially dangerous when used in a seismic damper. Yet, if NASA and the Defense Department standards were followed instead of ASTM, it becomes apparent that the material could only be used safely at much lower yield stress levels. In such cases, an engineer might be well inclined to select a much different material, in the interest of safety and professional liability. It would appear that using only ASTM standards, as they are presently written, gives the engineer virtually no guidance as to whether a material is suitable for any particular use. Other standards are much more robust and rigorous. Taylor Devices has had excellent success for more than 40 years using materials selected in accordance with the four standards organizations listed previously as 1-4, and uses these standards exclusively for seismic and wind dampers.

Examples of typical metallic materials used in the manufacture of dampers are provided in Figure 15.

PART	MATERIAL	SPECIFICATION
Piston Rod	15-5PH Stainless Steel, Wrought Bar, 160 ksi Minimum Yield Stress	MIL-HDBK-5 and AMS 5659
Cylinder	AISI 4340 Tubing, 120 ksi Minimum Yield Stress	MIL-HDBK-5
Piston Head	Bearing Bronze, Bar Stock, 50 ksi Minimum Yield Stress	AMS 4640
Seal Retainer	AISI 4340 Wrought Bar, 120 ksi Minimum Yield Stress	MIL-HDBK-5

FIGURE 15
METALLIC MATERIAL EXAMPLES

PERFORMANCE OF FLUID DAMPERS

A properly designed fluid damper will attenuate transient and steady-state inputs while staying within specified performance bounds. At the same time, the damper, as it is manufactured, must not yield, leak, or overheat during use.

Transient Inputs: Frequency and Response Time, Magnitude of Damping Needed

Depending on its end use, large dampers can easily be designed and constructed to attenuate input transients in the range of 0-2,000 Hz. However, for seismic applications, earthquake spectral inputs rarely contain much content at frequencies in excess of 10 Hz. Using conventional mechanical engineering practice for vibrating systems, a control device should be capable of operating at a frequency of at least 10 times the maximum input frequency. Thus, a frequency response range of 0-100 Hz is sufficient for most seismic damper applications.

Figure 16 provides the tested seismic transient response of a Taylor damper rated at 100 kips output. The results indicate that the damper can easily follow the input, and is behaving very predictably. This particular test was at the limit of the damper test machines available in 1995 that could generate transients, yet the damper itself is relatively small. Qualifications of damper suppliers should require that each manufacturer provide the transient response output of previously manufactured devices with output forces in the 50-100 kip range. This should be combined with previously published seismic transient test results on shake table testing of dampers installed in a representative scale building with multiple small dampers (usually in the 1-10 kip size range).

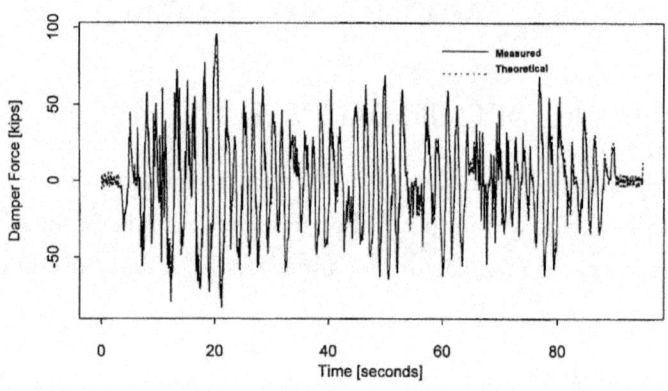

FIGURE 16

SEISMIC TRANSIENT RESPONSE OF A 100 KIP FLUID DAMPER

Impulse and frequency response of full scale dampers built for a specific project can readily be determined by drop testing, where a weight is allowed to free-fall a certain distance and then impact the damper. Figure 17 depicts the results of a drop test used to verify the high frequency and impulse response of a 150 kip output production damper. This particular test is loading the damper to 160 kips force. The test input was generated by dropping a 17,000 weight from 4.5 inches above the damper's piston rod. The opposite end of the damper was attached to a large seismic reaction mass of concrete and rebar.

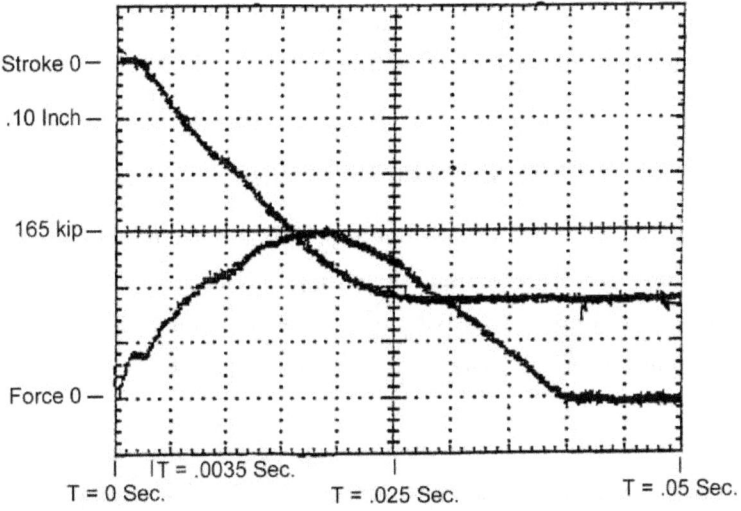

FIGURE 17 TRANSIENT DROP TEST RESULTS

For this particular test, the seismic reaction mass was approximately 1,500 tons, with a measured first mode resonant frequency of 970 Hz. The damper was designed for a base isolated structure, and had a total available stroke of plus or minus 23 inches. Instrumentation consisted of a load cell affixed to the end of the damper, and a stroke potentiometer located between the damper's piston rod and cylinder.

The test record was triggered by load cell response and the damper force can be seen to begin smoothly rising almost immediately after impact. The small horizontal shift in force from .001 to .003 seconds can be attributed to the drop weight rotating slightly and aligning itself to the damper piston rod, since stroking of the damper begins to occur shortly thereafter. Frequency response can be conservatively ascertained by determining the time from initiation of a load cell response to the time when both damper force and stroke are smoothly increasing. This is occurring on this test by the time T = .0035 seconds. Conservative frequency response would then be the reciprocal of time, or 285 cycles per second. Note the absence of any force "spikes" at impact, indicating that the dampers orifices begin flowing almost immediately. Any force spike at impact would constitute a test failure. This input and the damper's response substantiates the performance of a fluid damper in the presence of near fault seismic burst or the detonation of explosives.

Steady-state Inputs: Wind and Vibration

Structures in non-seismic regions normally use fluid dampers for the control or reduction of wind responses. Other applications for dampers relate to reduction in measured or felt vibration. This vibration can be initiated by various inputs, either internal or external to the structure.

Examples of externally induced vibrations include vehicular traffic, air traffic, or industrial sources. Internally induced vibration is usually caused by machinery sources inside the structure, or by sympathetic motions of the building's occupants, such as would be caused by large numbers of people dancing or marching. As noted previously, fluid dampers have been built to mitigate the response of inputs in the 0-2,000 Hz range. With respect to low amplitude vibrations, fluid dampers have been used to suppress amplitudes as low as .001 inch.

Qualification of damper suppliers for vibration inputs is the same as that required for seismic applications. Each damper manufacturer must provide representative output response data from previously manufactured devices, preferably in the 50-100 kip range. In the case of wind applications, it is usually possible to provide full scale damper response test data, recorded during transient testing with representative wind storm time histories.

Transient Inputs: Magnitude of Required Damping

The magnitude of fluid damping added to a structure for the suppression of seismic, wind, or other transient inputs is usually in the range of 5-45% of critical. This is a very wide range, and varies with the type of structure and excitation. Obviously the amount of damping selected is the responsibility of the engineer of record, but generalized damping levels from previous projects are tabulated in Figure 18 as guidelines.

If the specific structure is located at a soft soil site, then damping at the higher end of the ranges listed should be considered.

STRUCTURE TYPE	PERCENT OF ADDED DAMPING
Tall Buildings, Seismic and Wind Inputs	5-15%
Buildings 1-15 Floors, Seismic Inputs	15-25%
Bridges Non-suspension Type, Seismic Inputs	30-45%
Bridges Suspension Type, Seismic and Wind Inputs	15-25%

FIGURE 18
ADDED DAMPING LEVELS

Heating Effects

It is necessary for damper manufacturers to calculate the thermal response inside a damper to prevent overheating of internal parts during use. In most cases, overheating damage manifests itself by leakage, usually caused by a softened or melted dynamic seal. If calculations indicate that overheating is an issue, then in most cases the damper will be increased in physical envelope until temperature rise during operation is low enough so as to be safely accepted by the internal parts.

Thermodynamics teaches that there are three heat transfer processes, defined as convection, conduction, and radiation. In any given case, one of these processes will be dominant over the others. Without getting into very detailed mathematics, the following transport processes dominate in applications for fluid dampers:

Seismic and other short duration events: Conduction
Wind, steady-state vibration, long duration events: Convection and Conduction

It is very important that the correct transfer processes be known for a given application when the damper is being sized, and this is the responsibility of the damper manufacturer. There are numerous historical cases of inexperienced manufacturers providing products that do not have the required heat capacity for a given use. For example, the author remembers a competition for dampers to be used for recoil attenuation on a heavy machine gun. One inexperienced manufacturer arrived at the test with a damper having costly intricate cooling fins machined into the damper cylinder. Evidently the engineers at this firm assumed that machine gun dampers dissipate heat by steady-state conduction and convection, so cooling fins were needed. The other competitors were more knowledgeable about the application, and provided dampers of the same diameter, but with thick and heavy solid steel cylinder walls. The reason was simply that a machine gun will have the barrel of the weapon overheat dangerously if true continuous firing is attempted. Actual use involves firing 15-100 round bursts within less than 15 seconds, then letting the barrel cool. When the actual test was run, the damper with the wonderfully machined cooling fins suffered melted seals after only 1,000 rounds of testing. The dampers without cooling fins had no problems at all, because they were designed knowing that only conductive heat transfer was critical for the application - convection would not provide any substantive short-term transport of heat, since not enough time was available during a firing burst to establish convection. The balance of this section will explain why this phenomena occurs.

In the case of seismic and other short duration events, the damper mass must absorb virtually all of the input energy, by conducting the heat of energy dissipation to the damper's fluid and cylinder mass. Other parts are not involved, simply because the event ends before conduction to the end caps, piston rod, and mounting brackets can occur. No convection processes can set up with the air surrounding the damper, again simply because the event is of such short duration. Most damper manufacturers will allow a maximum credible seismic event to cause no more than a 100 degrees F heat rise over ambient temperature to occur. Measurement of the temperature is taken at the outside diameter of the damper cylinder, at a point axially aligned with the midpoint of the piston head location at the initiation of the transient. Because of the rather heavy construction of most fluid dampers, the peak temperature is not reached until several minutes after the input transient is applied. If a manufacturer has any seals located in the piston head itself, the allowable surface temperature *may be less* than the 100 degrees F over ambient value. This is because the energy dissipation occurs locally at the damper orifices, generating a localized "hot spot" inside the damper. This does not bother metallic parts or the fluid, which can accept temperatures of over 600 degrees F without problems. Seals located near these hot spots are an entirely different matter, since seal softening usually occurs when absolute temperatures exceed 300 degrees F.

When evaluating proposed damper designs for a seismic project, manufacturers will be fully capable of providing transient heating calculations for review. If desired, energy can be input to a full-sized damper by means of sine wave cycling until the energy input from the test equals the integrated energy from a maximum credible transient. Oddly enough, the actual number of full-force, full-stroke damper cycles to equal the energy of a real seismic transient is quite small. Figure 19 provides a tabular list of transients, and the equivalent energy number of full sine wave cycles at the design level maximum velocity. The calculations assumed that the damper was designed with a 1.5:1 safety factor on required minimum stroke, which is standard practice for seismic dampers.

Projecting the data in Figure 19 to the proverbial "big one" in California can be done using a method suggested by Housner [5]. The result indicates that the El Centro earthquake scaled to 0.5 g would be equivalent to a M = 8 - 8.5 event, and is equivalent in energy to 1.62 sine wave cycles.

SEISMIC INPUT	EQUIVALENT NUMBER OF FULL DAMPER CYCLES
El Centro, Imperial Valley, 0.35 g S00E	.79
Pacoima Dam, S74W	.52
Taft, Lincoln School Tunnel, S69E	.45
Northridge, 90 Degrees	.34
Kobe, Japan, E-W	.32

FIGURE 19 EQUIVALENT SINE WAVE CYCLES

Damper heat transfer for wind or steady-state vibration applications uses completely different heat transport methods from short duration transients. The reason is simply that most wind storms or steady-state vibrations are assumed to last at least several hours, with the damper being continuously cycled. Tests have indicated that a damper under continuous cycling inputs will reach a near steady-state temperature within the first ½ hour of cycling, and at that point most heat transfer will occur by convection to the air surrounding the damper. In most cases, convective air currents will distribute the damper's heat throughout the available air mass inside the structure. In some cases, the engineer of record will provide air ducting to the dampers, preferably using outside air driven by the wind itself. Heat transfer calculations for wind dampers are relatively complex, and the damper manufacturer must be given generalized wind motion data to properly size the damper. In general, manufacturers allow steady-state heating of the damper to be no more than 100 degrees F over ambient, similar to the allowable heat rise for seismic use. Applications of dampers sized for seismic zone applications which also will see wind deflections are relatively common today. In nearly all cases, the damper size is governed by the large seismic inputs; wind inputs being relatively of little consequence, when the damper is large to begin with. A second consideration is that the optimal amount of critical damping for seismic applications is usually much greater than that required for wind applications. The additional damping serves to provide excellent occupant comfort, and also serves to greatly reduce the amount of power each damper must dissipate. This is easily understood when one considers that wind motions involve some amounts of resonant or quasi-resonant structural motion. As was noted earlier, the magnification factor for a structure under forced resonance is:

$$\text{magnification factor} = \frac{1}{2\zeta}$$

$$\text{where } \zeta = \text{the damping ratio } \frac{c}{c_{cr}}$$

Thus, if the damping ratio doubles, the resonant magnification reduces by a factor of 4. Since the magnification factor is directly proportional to the wind power the damper must accept, the relative power a damper must dissipate rapidly diminishes as the damping ratio increases.

Cyclic Life - Service Life

A properly designed and manufactured damper should not require any type of periodic service. This is simply because if the proper seal is selected by the manufacturer, the damper will be essentially "dry sealed", with high seal scraping forces used to eliminate any static seal weepage. In the early years of dampers, most seals were produced for use in hydraulic systems, where hydraulic cylinders are used to perform work. Since a hydraulic cylinder is expected to promptly and accurately move to specific positions, even small amounts of seal friction will degrade the resolution of the hydraulic system. Thus, most hydraulic systems use dynamic seals that are intended to continuously leak, both statically and dynamically. Since dampers are passive elements, system resolution is not a design parameter, so each damper manufacturer has developed proprietary dry seals to prevent any measurable leakage during service. Sometimes, manufacturers with little or no damper experience attempt to make dampers from commercial hydraulic cylinders. Usually, these products are easily discovered, since fluid level sight glasses or external oil tanks will be observed, with the associated plumbing. In other cases, the damper=s warranty will require that periodic servicing, oil changes, or even seal changes are required.

The type of dynamic seals used in dampers are limited in life by wearing of the seal as the piston rod moves back and forth. In general, seal life is measured in terms of the total number of inches of rod displacement during a damper's lifetime. Present day seal designs are considered as so robust that a well-built damper should be warranted by the manufacturer for at least 35 years. The warranty should state that no periodic fluid replenishment or periodic servicing of any type is required.

Fire Resistance

In general, dampers are fabricated mostly from steel, utilize internal fluids from the silicone family, and contain the fluid with polymer or elastomer seals. In actual service, the dampers are not a part of the primary load bearing path of the building, since a damper does not produce or accept static forces. Unfortunately, this has served to place dampers in the "gray area" as to how fire rating procedures should be applied to this component.

The failure mode of a damper under fire conditions is relatively benign, i.e., the seals soften and eventually melt, causing loss of fluid. Dampers that use external gas charged accumulators raise the safety issue of the

DAMPERS: HOW VISCOUS DAMPERS PROTECT STRUCTURES

accumulator itself, which is usually classified as a gas bottle, and therefore is subject to severe scrutiny by the fire codes.

Depending on the specific placement and location, all of the following approaches have been used on previous projects:

1. Consider the dampers as a non-structural component, and take no special provisions.
2. Use sprinkler systems nearby, as are used presently for the protection of rubber seismic base isolation bearings.
3. Spray the exterior surfaces of the damper with fire retarding paint.
4. Box the dampers inside interior spaces with sheetrock.
5. Perform calculations to verify a specific fire rating.

In the event that number 5 is selected, the typical fire rating for an unpainted damper will be in the range of 1- 3 hours, with larger diameter dampers having the higher values.

Customer Controlled Parameters

Fluid dampers for a specific project are essentially adjusted by the manufacturer to meet specific customer specified parameters. The parameters include:

1. Maximum rated force
2. Minimum safety factors to yield
3. Minimum required useable deflection from neutral position
4. Damping constant
5. Damping exponent
6. Operating temperature
7. Maximum wind power input (if applicable)
8. Maximum damper envelope
9. Damping mounting configuration

The maximum rated force of the damper is usually the force expected during the maximum credible event that the device is designed for. The safety factor to yield is based upon either the maximum rated force, or the velocity at which this maximum force occurs. Typically, the safety factor is 1.5 to 1, meaning that the damper will not yield when subjected to a force or velocity 150% of the rated maximum. Structural engineers seem divided equally as to whether a force or a velocity basis is used. In general, rating at 150% velocity gives a uniform safety margin for any project. Rating on force alone may give an excessively large or small safety factor, depending on the dampers exponent. For example, a V^2 damper will have its load increase to 150% of rated load at a velocity increase of only 22% from maximum. Conversely, a $V^{.3}$ damper would have to be operated at almost 400% maximum speed before its force would increase to 150% of maximum.

The damping constant, damping exponent, and temperature ranges can be easily expressed on a graph, defining allowable damper performance bandwidth at any defined operating temperature. Figures 20 and 21 provide damper performance graphs for two separate projects.

Figure 20 defines damper performance for a 250 kip force linear damper used in diagonal brace elements of a steel moment frame building. A nominal function is the middle line in the plot, with high and low tolerance limits applied. Operating temperature range is +32 degrees F to +120 degrees F, and the damper's output must fall within the plus or minus 20% tolerance at all velocities, at any defined operating ambient temperature, both before and after the damper has absorbed the energy of a maximum credible seismic event. Note the deviance allowed in the band at very low speeds. This is done in deferral to damper testing apparatus, where large load cells, stroke potentiometers and large test machines have trouble reading very accurately at low speeds.

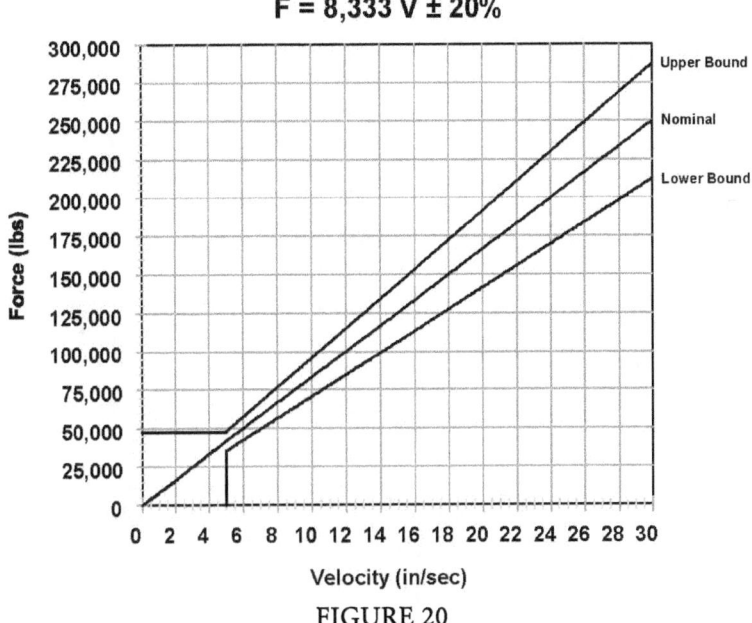

FIGURE 20
DAMPER PERFORMANCE BANDS FOR A LINEAR FLUID DAMPER

Figure 21 depicts required performance for a large base isolation system damper, with the high velocities associated with Zone 4 seismic regions of the United States.

The structural engineer wanted to hold a narrow plus or minus 15% tolerance band over a +32 degrees F to +130 degrees F temperature range. Again, a larger acceptance "window" is used at low velocities to accommodate available test equipment. This particular damper is rated at 310 kip force, and has a non-linear $V^{.4}$ damping characteristic.

Specific applications may impose diameter or length restrictions on the damper, and these maximum values can also be customer specified. In most cases, a diameter limitation is much more common than a length restriction.

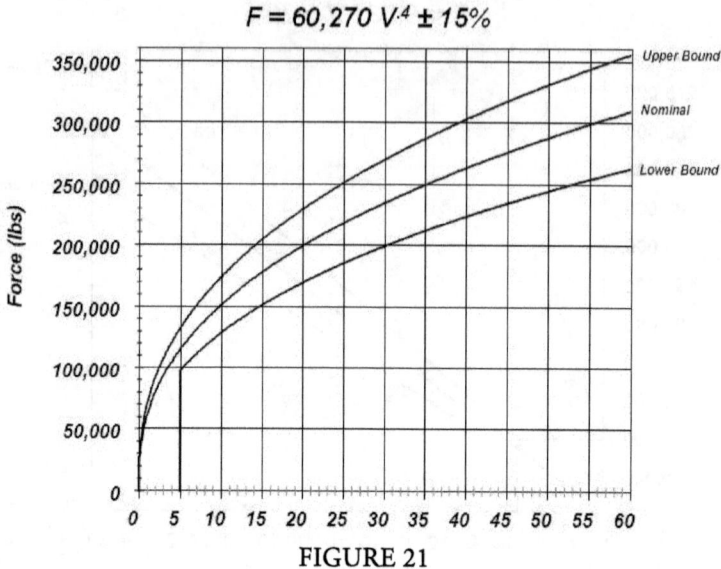

FIGURE 21

DAMPER PERFORMANCE BANDS FOR A LOW EXPONENT DAMPER

IMPLEMENTATION OF FLUID DAMPERS

One of the most beneficial aspects of using fluid dampers in a structure is that they are essentially a "bolt-in" item, of a relatively compact size. If used as part of a structural bracing system, the fluid dampers usually will have a smaller cross-sectional envelope than a conventional steel brace.

Fabrication Issues: Size vs. Cost

If a given structure requires certain total macroscopic damping, to implement this damping will involve dividing the total damping by the number of dampers used. The end result is a maximum force and damping function for each individual damper. The question is: Should the engineer select a large number of small dampers, or a lesser number of large dampers? The rather large number of available dampers sizes tends to compound the problem even further.

The structural engineer normally starts out with multiple dampers of the same size, dispersed uniformly throughout the structure. This usually results in many dampers in the relatively small force range of 20 kip to 50 kip output. If the structure is small enough to require less than 32 pieces of a 20 kip to 50 kip damper, than this will probably be the most effective size, since quantities smaller than 32 pieces tend to become costly, due to set-up, engineering, and test charges being amortized over a small quantity. The 32 piece number was obtained strictly from the past experience of the author.

The next step is to reduce the number of dampers by using the next larger size, and continuing this process until:

1. The quantity of dampers goes below 32 pieces.
2. The force rating of the damper goes over 600 kip.
3. The structure begins behaving less efficiently because the dampers are not distributed well enough.

This is definitely an interactive process, and thus far has proven to have a great deal of variance from project to project.

Currently available damper sizes from present manufacturers are:

30 kip	200 kip	500 kip	1,000 kip	2,000 kip
50 kip	300 kip	600 kip	1,250 kip	
100 kip	400 kip	750 kip	1,500 kip	

In terms of relative cost, the least expensive sizes on a force basis are in the 300-600 kip range, i.e., one piece of a 300 kip damper costs less than 10 pieces of a 30 kip damper. In most cases, dampers larger than 1,250 kip are used only on large bridges, since the point loading into a building structure from such a large device requires special design considerations be made to the structure's beam to column connections. Also, note that just as 30 kip dampers are relatively expensive compared to the 300 kip size, dampers larger than 600 kip also tend to become costly. In this case, the problem relates to a general lack of available high-strength steel in the very large sizes, requiring special orders to the steel mill.

Tuned Mass Fluid Dampers for Tall Buildings

A special design case for fluid dampers occurs when they are selected for use as part of a tuned mass damper system for tall buildings. This application suspends a large internal lumped mass at the uppermost floors of a tall building, supporting the mass with cables, steel arms, or springs combined with air/fluid/mechanical slider bearings. The end result is to have the mass centered within the building on lateral spring elements. The dampers, which typically have strokes in the 2-6 foot range, are used to control the response of the tuned mass and springs. When the building is subjected to wind inputs, the building will tend to move with the wind. The tuned mass, due to its inertia and long period attachment, prefers not to move. Thus, the tuned mass reduces deflection of the building under wind inputs by essentially applying force to the building in a direction resistive to the wind motion. Tuned masses are not used to provide seismic protection, since they are both ineffective and even dangerous under the high energy and unpredictability of a seismic event.

The biggest problem encountered with tuned mass dampers is that they must stroke long distances, almost continually, for the life of the building. This naturally brings up the issue of seal wear, which in itself is compounded by the tuned mass damper requiring very low seal friction to allow the mass to move freely. The end result is that a conventional damper modified with low tension seals and used in a tuned mass system will require that seals be replaced every 1-2 years.

An attractive alternate to this problematic application is to use a damper that does not have sliding seals. To achieve this design configuration, all points in the damper where a sliding seal would be placed have the seals replaced with roller bearings. Next, flexible metal bellows are attached over the piston rod to retain fluid. The use of flexible metal bellows seals comes from the U.S. space program, where the use of hydraulic components in the vacuum environment of outer space requires that fluid be hermetically sealed at all times, to prevent out-gassing (in outer space, fluid materials exposed to vacuum turn into a gaseous cloud, which can interfere with the operation of electronics or optics). Flexible metal bellows are made from multiple leaves of stainless steel sheet, edge welded to make a bellows-like shape. Any desired damper stroke can be accommodated, simply by welding on more leaves.

Because the metal bellows seals by flexing, a damper using this type of seal has highly desirable near-zero friction, combined with a leak free life up to 500 million cycles. The only disadvantage is that the metal bellows are extremely labor intensive, and usually cost more themselves than an entire damper with conventional seals.

Linkage Driven Dampers for Stiff Structural Systems

Structures that are inherently stiff or rigid often require that fluid dampers function with extremely small displacements, down to plus or minus 0.05 inches or less. Unfortunately, when a damper is operated with small displacements, relatively small damping pressures are generated. This is because all fluids are slightly compressible, and the cylinder of the damper expands slightly under pressure.

Two solutions are presently available to address this problem. One is simply to use a very large damper with heavy cylinder walls, operating at very low pressures. This solution is costly, and dampers operating at very low pressures are subject to viscous effects, changing their output force radically as the ambient temperature varies the viscosity of the damping fluid.

The second solution is more practical, and comes from the field of mechanical engineering. This solution involves the use of a lever driven mechanism to multiply the deflection of the structure. Figure 22 depicts such a mechanism, called a toggle brace element. Shake table tests on this mechanism by Taylor and Constantinou [6] revealed that it was possible to obtain a 3:1 increase in the measured story drift at the mounting points of the damper. Overall performance of the tested system was virtually identical to that obtainable with a larger and much more costly direct acting damper, without encountering the viscous effects problem.

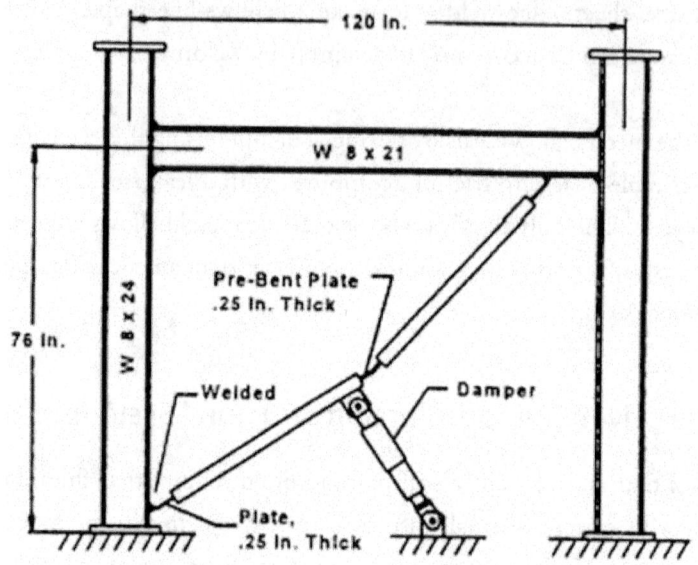

FIGURE 22

TESTED FRAME WITH TOGGLE BRACE-DAMPER SYSTEM

Full Scale Testing Issues

Testing of large fluid viscous dampers is somewhat difficult; for all intents and purposes, the test machine has to produce the equivalent of a real earthquake. For example, the large damper shown in Figure 10 requires 3,000 horsepower (2.1 megawatts) to drive it at full velocity, requiring tremendous testing capabilities to test just a single damper.

In view of this problem, test methodology has evolved within the structural engineering community to allow fully tested dampers to be implemented into structures without excessive cost. The methodology involves a four step approach, the first two steps having been discussed previously in the section "Performance of Fluid Dampers". The next two steps apply to full scale dampers, manufactured under contract for actual use in a structure.

The four steps are:

First, the manufacturer must qualify and fully validate the damper's generic design through shake table tests with small dampers installed in a building model.

Second, test dampers in the 50 kip - 100 kip output range must be tested to verify frequency response under seismic and/or wind transient inputs, in addition to temperature drift.

Third, the first two production dampers produced will have the following tests performed:

1. Force vs. velocity tests to 125% maximum rated force with sufficient data points taken to verify that the damping functional output is to specification.

    Test success criteria: All data points within designated limits of the specification.

2. Temperature testing, same loads and data points as in number one above, but test at low and high ambient temperatures per specification.

    Test success criteria: All data points within designated limits of the specification.

3. Thermal energy capacity, seismic: An energy equivalent to that found by integrating the maximum credible transient shall be input to the damper, with the damper insulated via thermal blankets to prevent heat loss. The damper shall then be tested, same loads and data points as in number one above.

    Test success criteria:

1. All data points within designated limit of the specification.
2. No visible drop-wise leakage or parts breakage observed.

4. Thermal energy capacity, wind: Dampers used only for wind energy dissipation do not normally require seismic testing per 3 above. Instead, the damper must be cycled at a displacement and velocity such that is dissipating the required wind power on a steady-state basis. In cases where the power must be continuously dissipated for several hours, testing should take place with an external environment similar to the actual end use.

Test instrumentation for this test should be such that damper output force, displacement and velocity are monitored, such that the required power dissipation level is obtained within specified parameters. Damper temperature should be constantly monitored, with temperature transducers typically located on the cylinder of the damper itself, with additional temperature data recorded for the air surrounding the damper.

Test success criteria:

1. Specified power dissipation to be continuously provided by the damper.
2. Damper force, displacement and velocity data points to be within designated limits.
3. No visible drop-wise leakage or parts breakage observed.
4. Thermal energy capacity, wind: Dampers used only for wind energy dissipation do not normally require seismic testing per 3 above. Instead, the damper must be cycled at a displacement and velocity such that is dissipating the required wind power on a steady-state basis. In cases where the power must be continuously dissipated for several hours, testing should take place with an external environment similar to the actual end use.

Test instrumentation for this test should be such that damper output force, displacement and velocity are monitored, such that the required power dissipation level is obtained within specified parameters. Damper temperature should be constantly monitored, with temperature transducers typically located on the cylinder of the damper itself, with additional temperature data recorded for the air surrounding the damper.

Test success criteria:

1. Specified power dissipation to be continuously provided by the damper
2. Damper force, displacement and velocity data points to be within designated limits.
3. No visible drop-wise leakage or parts breakage observed.

Fourth, all dampers shipped to the site, including the first two units from step three above, shall be subjected to the following tests prior to shipment:

1. Proof pressure test: Pressurize unit to 125% maximum operating pressure. Hold pressure for 3 minutes minimum, then release pressure.

   Test success criteria: No leaks or parts breakage observed.

2. Force vs. velocity tests: Test damper in each direction obtaining maximum operating force defined by specification.

   Test success criteria: The velocity at which maximum force is obtained shall produce a data point within designated limits of the specification.

3. Measure and record available stroke of damper.

   Test success criteria: Verify that stroke is within specification limits.

4. Measure all damper dimensions specified on the project drawings.

   Test success criteria: All dimensions to fall within specification values.

Note that each and every damper delivered to the job site should be tested to verify force capability, damping function, and measurable dimensions. This is done in view of the critical importance of seismic dampers to a structure's performance, and the potentially litigious nature of owners. The testing of all dampers eliminates many potential product and professional liability issues for this type of product.

The cost of this testing is quite small, considering that much of the testing costs are incurred during the set-up and rigging of the test machine. Also note that no attempt is made to specify a particular type of test machine for production damper testing. Some manufacturers prefer to test with hydraulic actuators, some prefer to drop test, and some use a combination of the two methods. Additional information on the testing of large fluid dampers for seismic energy dissipation is provided by Taylor and Constantinou [7].

At the time this document was prepared, general guidelines for testing of all types of seismic and wind dampers were being prepared by the National Institute of Standards and Technology (NIST). This organization is part of the United States Department of Commerce, and the guidelines are being prepared by NIST's Building and Fire Research Laboratory. The intent of the guidelines is to facilitate development of damping technology while giving the end user a high level of confidence in the installed devices.

Detailing Issues: Attachments and Brace Styles

There are only three basic ways to attach dampers into a building or bridge structure.

DAMPERS: HOW VISCOUS DAMPERS PROTECT STRUCTURES

1. Base isolation dampers have clevises and spherical bearings at each end. These long stroke dampers are connected to the foundation and to the building frame respectively, using mounting pins. Note that the mounting pins for base isolation dampers *must* be oriented vertically, to allow proper articulation during out of plane motion.

2. Dampers for chevron bracing systems have clevises and spherical bearings at each end. Connections are similar to base isolation dampers, except that the mounting pins are usually oriented horizontally. The typical plus or minus 5 degree rotation angle of a spherical bearing will accommodate out of plane motion for the relatively small drifts encountered with this type of installation.

3. Dampers for diagonal bracing systems have a clevis with spherical bearing at one end, and a mounting plate at the opposite end. The mounting plate attaches to a brace extender.

Schematics of the three basic mounting attachment styles are provided in Figure 23. The mounting pins used to attach the dampers to brackets or tangs are often supplied by the damper manufacturer. In most cases, the manufacturer will also provide the brackets or tangs used to connect to the building structure. The reason for this is that the pins must be fitted very closely to the clevises and spherical bearings, to insure that the connection has no discernible play. The pins themselves are heavily loaded, and are normally fabricated from 120 ksi yield or higher stainless steel. Each mounting pin extends entirely through clevises and tangs, and includes cross drilled holes at each end. Large cotter keys are inserted into the cross drilled holes to finish the installation. Figure 24 depicts a spherical bearing and Figure 25 is a schematic of a completed attachment.

The spherical bearings are installed into the damper clevis by the manufacturer. Depending on the structure, the bearings themselves can be plated steel, coated steel, or stainless steel. For bridge applications, seals are added to stainless steel or plated steel bearings to reduce potential for corrosion.

There is a certain degree of variance in the design of diagonal brace elements. In many cases, the building's out of plane motion is small enough that a long extender brace can be welded to the beam to column connections, with the opposite end of the extender attached to the damper. Thus, the damper's single clevis and bearing is the only connection capable of articulation. If the brace and damper elements are relatively short, then clevises must be used at each end of the damper-brace combination.

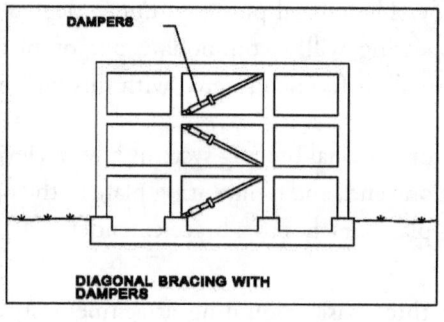

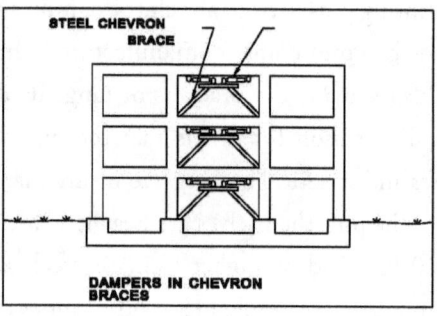

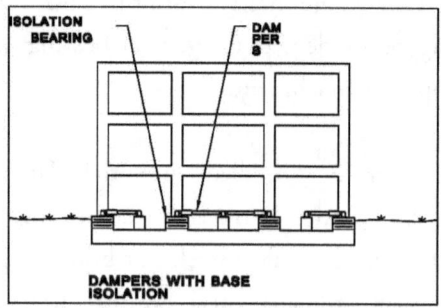

FIGURE 23

BASIC MOUNTING ATTACHMENT STYLES

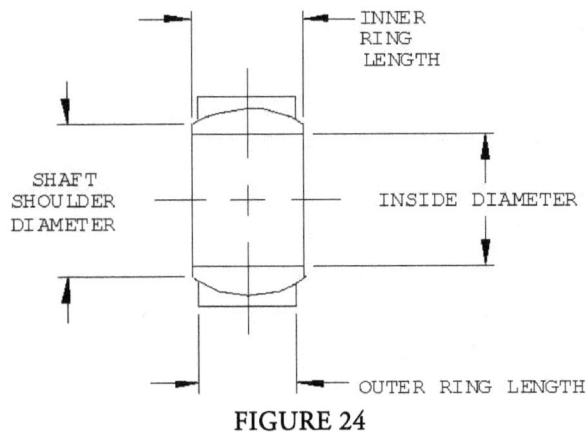

FIGURE 24
SPHERICAL BEARING

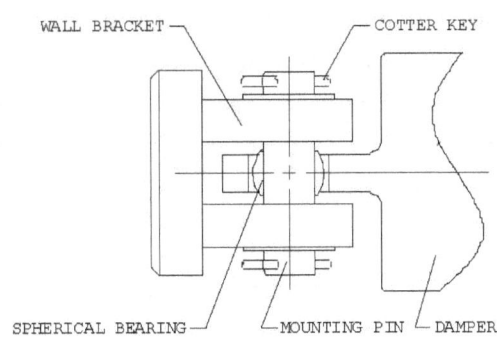

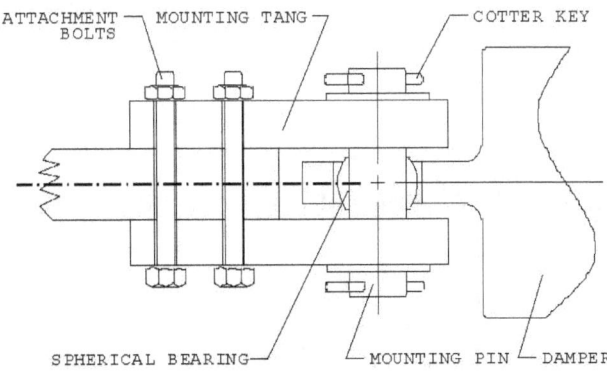

FIGURE 25
SCHEMATIC OF A COMPLETED ATTACHMENT

Maintenance and Inspection

Maintenance is not required for a properly designed and manufactured fluid damper used for seismic and wind damping in structures. Usually, visual inspection of the dampers should occur after a major seismic or wind storm event. In general, visual inspection involves looking for discernible leakage or broken parts and/or connections. In the event of seismic overload, the damper mounting pins may bend or shear. Pin condition can be ascertained by simply rotating the pin, checking to see if the damper clevis "wobbles" as the pin is turned.

After a major seismic event, some structures may require an enhanced inspection, due to regional code requirements. This may involve as little as verifying that the damper is full of fluid. In some cases, regulations may require that a few dampers be removed at random from the structure, and subject to testing to verify damping output. In all cases, any site personnel who will be involved in repairing dampers or replacing damper parts must be employees of the damper manufacturer. This is necessary for insurance and product liability considerations.

PROJECT EXAMPLES

At this time, more than 700 major structures are using fluid dampers to obtain enhanced performance during seismic or wind excitation. Four of these projects are described here, selected specifically for their diversity. A portfolio of photographs for these projects is provided following the Conclusions section.

The Arrowhead Regional Medical Center at Colton, California

This project was the first application for fluid dampers in the seismic protection field. The five buildings of this complex use a total of 186 dampers, each being rated at 320,000 lb. force.

The dampers are used to dissipate seismic energy, and are installed in "systems parallel" with rubber base isolation bearings. The 850,000 square foot medical center is located in San Bernardino County, between the cities of Ontario and San Bernardino. The location is within 8 km of the San Andreas Fault, and 10 km of an intersecting fault. The location of the medical center was determined by available Federal Government funding, provided with a requirement that the complex be located between the two cities noted, and readily accessible from the I-10 Freeway.

The original design concept utilized only the base isolation bearings. However, the near fault location generated site transients in the 60 in/sec. range, requiring that the bearings provide deflection in excess of plus or minus five feet! No rubber bearing of this size had ever been fabricated. Of course, the large deflection also required that new technology be created with respect to utility connections, parking ramps, access walkways, etc.

In an attempt to reduce deflections, the project design team investigated adding 37% fluid damping to the 8% damping expected from the so-called "high damping" rubber bearings, for a total damping level of 45% critical. Higher damping levels were found to cause decreased deflections, but higher stress, so the 45% damping level was considered as optimal for both stress and deflection. The use of the fluid dampers allowed bearing displacement to be reduced to only plus or minus two feet, a reduction so substantial that the project goals could be accomplished economically. One of these goals was to produce the most seismic resistant civilian structure in the world, and this was easily achieved with fluid dampers.

FIGURE 26
ARROWHEAD REGIONAL MEDICAL CENTER

FIGURE 27
ARROWHEAD REGIONAL MEDICAL CENTER INSTALLATION

FIGURE 28
ARROWHEAD REGIONAL MEDICAL CENTER CLOSEUP OF ATTACHMENT OF DAMPER

The San Francisco Civic Center Office Building

Those who are familiar with seismic designs in San Francisco will agree that this historic city is literally the home of the braced steel seismic resistant frame. When the 1994 Northridge, California earthquake revealed problems with steel moment frames, one would normally anticipate the desire for braced frames to become even stronger. One can then imagine the surprise within the structural engineering community during 1996, when erection began on the 14-story San Francisco Civic Center office building. This 800,000 square foot structure combined 292 fluid dampers with the so-called post-Northridge moment frame to optimize performance while maintaining a cost-effective project budget. Two different damper force levels were used by the engineer, 225 kip and 125 kip. All dampers were plus or minus 4 inches stroke. The dampers were used in diagonal brace elements, with a bolted flange connection to attach dampers to their brace extenders. The piston rod end of the damper incorporated a clevis with spherical bearing, as did the opposite end of the brace extenders. The dampers were supplied with building attachment clevises, which consisted of simple tang plates that were bolted to a gusset plate at the building=s beam to column connections. Low exponent damping in the $V^{.4}$ range was selected, combined with a building frame that can provide extensive inelastic deformation. The use of low exponent damping tended to limit damper forces when the frame was loaded into the inelastic range.

FIGURE 29
SAN FRANCISCO CIVIC CENTER DAMPERS
READIED FOR FINAL PAINTING

FIGURE 30
SAN FRANCISCO CIVIC CENTER DAMPER INSTALLATION

DAMPERS: HOW VISCOUS DAMPERS PROTECT STRUCTURES

FIGURE 31
SAN FRANCISCO CIVIC CENTER
DAMPER CLOSEUP SHOWING CONNECTION DETAILS

The Hotel Woodland Woodland, California

This 4-story, 1928 vintage structure is in the Federal Registry of National Historic Landmarks. The building is constructed from non-ductile reinforced concrete, consisting of an RC frame at the first level, and RC shear walls at levels 2, 3, and 4. This type of structure is generally identified as having a "soft first story." The owner wished to convert the building into office space, and elected to upgrade the seismic resistance. A simplistic approach would be to add external shear walls, but of course this would essentially eliminate the historic exterior of the building.

An extensive structural analysis was performed, looking at potential improvement areas. The most promising solution was to add fluid dampers to the first story of the structure. To accomplish this, steel moment frames were added to the first story to provide stiffness and strength, which the existing lightly reinforced concrete columns did not have.

Chevron bracing was added to install the dampers, allowing damper loads to be limited to the horizontal plane only, the dampers being mounted parallel to the floors. A total of 16 dampers was used, with two dampers in each chevron brace. Each damper was rated at 100 kip output, with available deflection of plus or minus two inches.

FIGURE 32
HOTEL WOODLAND

FIGURE 33
HOTEL WOODLAND - INSTALLATION OF DAMPERS

The Pacific Northwest Baseball Stadium Seattle, Washington

This major league baseball park opened during the 1999 season, and features a three-section retractable roof, of steel truss construction. When fully extended, the roof measures 630 ft. x 655 ft., is 210 ft. in height, and weighs 11,000 tons.

Potential inputs to the roof include Zone 4 seismic transients and high winds. Added fluid damping was selected at an early stage of the design process to reduce lateral seismic and wind loads to the roof. The reduced loadings from the damped structure reportedly provided a net savings of $4.2 million on the project. Additional dampers were added between roof sections to eliminate the potential for longitudinal pounding damage. The latter application proved mundane, using virtually off-shelf dampers in the 200 kip B 400 kip range. The use of dampers in the lateral direction proved much more difficult, since the only available mounting point was to use large dampers in diagonal braces between the column and roof trusses. The dampers were located relatively close to the intersection of these two structural elements, and available mounting regions dictated that a total of only eight dampers could be used. These eight were required to reduce both stress and deflection in the 11,000 ton roof, requiring that each damper be capable of 1,100 kip output under maximum credible earthquake conditions, with plus or minus 15 inches deflection. The design was made much more difficult due to the 23 feet required pin to pin length for the dampers, coupled with a restriction from the architect that would not permit a conventional flange connection between damper and extender. The damper design was further restricted by the architect with a requirement that there be no more than a 25% diameter change at any point along the entire length of the device.

A third restriction on this design was imposed by wildlife in the local environment, namely seagulls and other birds, which would be expected to inhabit the roof truss structure. Seagulls are notorious for having the capability of ingesting practically anything that resembles food, and have evolved a powerful digestive system. Experiments have verified that gull droppings can etch or remove most paints, and can even cause erosion to plated steel. The problem was compounded by the fact that seagulls find shiny surfaces to be attractive, and this is especially true of highly finished piston rods. To address the concern, the large dampers for this project used steel covers over stainless piston rods. The rod covers were combined with a forged stainless steel cylinder which was passivated after machining and heat treatment. Passivation is a chemical process that removes any free iron atoms from the surface of the steel, rendering the steel much more corrosion resistant than plain stainless steel. Passivation also "knocks down" much of the shiny, reflective nature of stainless steel surfaces, leaving a dull finish that will hopefully prove unattractive to birds. The project painting requirements specified a relatively thick coating of premium paint, in an attempt to provide enhanced resistance. The actual damper installation proved uneventful, and so far the dampers seem unaffected by the local fauna.

FIGURE 34
PACIFIC NORTHWEST BASEBALL STADIUM

DAMPERS: HOW VISCOUS DAMPERS PROTECT STRUCTURES

FIGURE 35
TESTING ON A DAMPER CARTRIDGE FOR THE PACIFIC NORTHWEST BASEBALL STADIUM

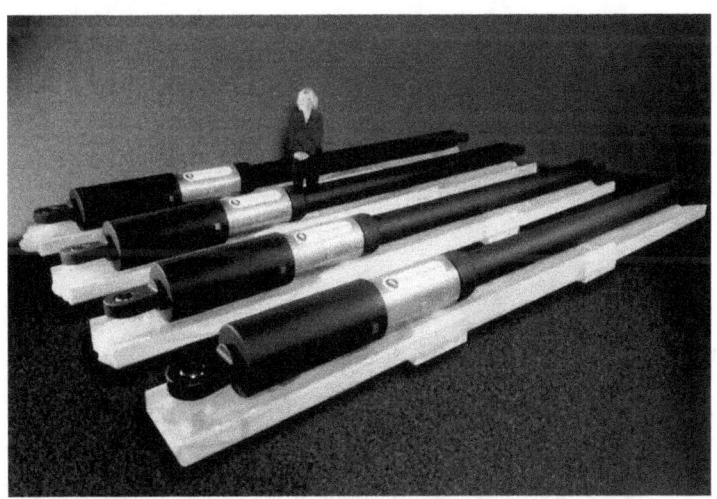

FIGURE 36 COMPLETED DAMPERS FOR THE PACIFIC NORTHWEST BASEBALL STADIUM

FIGURE 37
DAMPER INSTALLATION AT THE
PACIFIC NORTHWEST BASEBALL STADIUM

CONCLUSIONS

The use of fluid dampers for seismic and wind protection of commercial and public structures has occurred widely throughout the 1990's. Implementation has occurred rapidly, compared with other technologies. This is due largely to the widespread use of these products on Cold War era defense and military programs. When the Cold War ended, much of the fluid damper technology was declassified and transitioned to the public for commercial use. Very little development was needed to put dampers to work in civil engineering structures, simply because substantial taxpayer dollars had already been spent throughout the Cold War developing optimal damper designs. These were proven through extensive testing and widespread use throughout the military and defense sector.

When fluid dampers are used for seismic or wind protection, the end result is a predictable reduction of both stress and deflection in the structure. Indeed, this simultaneous stress and deflection reduction is unique to fluid dampers. Optimal performance is dependent on the type of structure and the level of performance required. Damping levels for optimal use of this technology range from 10% critical to 45% critical.

Today, more than 700 major buildings and bridges are using fluid dampers as a primary design element. Damper sizes being used range from as little as 30,000 lb. force to more than 1.5 million lb. force, with deflections as low as one inch and as high as 60 inches. Indeed, it can be said that the use of supplemental fluid dampers will be one of the primary solutions for seismic and wind protection in the structures of the 21st century, or in the more succinct form quoted by one California structural engineer:

WHEN IN DOUBT, DAMP IT OUT!

REFERENCES

1. Thomson, William, 1965, *Vibration Theory and Applications*, Prentice-Hall, Englewood, Cliffs, New Jersey.

2. NEHRP Commentary on the Guidelines for the Seismic Rehabilitation of Buildings (FEMA 274), April 1997, prepared by Applied Technology Council (ATC-33 Project), Redwood City, California.

3. Constantinou, M.C., Symans, M.D., 1992, "Experimental and Analytical Investigation of Seismic Response of Structures with Supplemental Fluid Viscous Dampers," Technical Report NCEER-92- 0032, National Center for Earthquake Engineering Research, Buffalo, New York.

4. Hogg, I.V., 1971, *The Guns 1914-1918*, Ballantine Books, Inc., New York, New York.

5. Housner, G.W., 1965, "Intensity of Ground Shaking Near the Causative Fault," *Proceedings of the Third World Conference on Earthquake Engineering*, Vol. 1, New Zealand.

6. Taylor, D.P., Constantinou, M.C., 1998, "Development and Testing of an Improved Fluid Damper Configuration for Structures having High Rigidity," *Proceedings of the 69th Shock and Vibration Symposium*.

7. Taylor, D.P., Constantinou, M.C., 1995, "Testing Procedures for High Output Fluid Viscous Dampers used in Building and Bridge Structures to Dissipate Seismic Energy," *Journal of Shock and Vibration*, Volume 2, Number 5, John Wiley & Sons, Inc., New York, New York.

Introduction to Part 2

Taylor Devices was founded by Paul Taylor in 1955. Its first product was a liquid spring, followed by a number of shock absorbers and fluid dampers. Gross income for Taylor Devices was five million a year at most for many years. Then Paul's son Doug Taylor took over in 1990.

Doug Taylor introduced a number of new products, including a line of seismic protection dampers that were very successful. Gross sales increased to 35 million a year over the years. Doug's dynamic leadership solidified the company into the leading supplier of fluid dampers in the world.

This is the story of Doug Taylor's relationship with his father, his many experiences with fluid dampers, and how his leadership propelled the company into world leader.

DOUGLAS P. TAYLOR Taylor Devices, Inc.,
North Tonawanda, New York, USA

SUMMARY

This article profiles my life and career highlights. I am an engineer who specializes in the mitigation of shock and vibration for structures of all types; from buildings and bridges to aircraft, naval vessels, and spacecraft. It is the story of how I entered the profession during the Cold War period and went on from there to invent many damping concepts. In some cases, names and project highlights have been omitted for security reasons. This is the nature of some of my work.

1. INTRODUCTION

First an admission. My degree is in Mechanical Engineering, not Structural Engineering. I graduated from the State University of New York at Buffalo with a Mechanical Design Engineering degree, with minors in shock and vibration control, and that nearly lost engineering art, metallurgy. As a result, nearly my entire career has been spent working with structures. Many structural engineers believe that mechanical engineers always design machines or HVAC systems, and this is indeed true of many of us. When I received my degree in 1971, I was presented with a very heavy-duty book by the ASME, a book by Babock and Wilcox entitled Steam. I skimmed the book and realized that this was a book that might well be cherished by traditional mechanical engineers—but certainly not by me. I eventually used the book exactly one time, to get information about the design of nuclear power plants in the pre-containment dome period. I needed this information for two of my good customers: one who wanted me to figure out how large a warhead was needed on an ICBM to damage a plant of this type beyond all hope of repair. The second customer wanted me to design an isolation system to allow a plant of this type to remain operational if a warhead of X kilotons was detonated at ground level Z feet from the plant. My story begins with what I have termed 'The Sandbox Incident' and ends with discussions about the use of damping devices to protect buildings and bridges from seismic events. As the Grateful Dead sang while 'Shuffling off to Buffalo,' 'what a long strange trip it's been . . .'

2. THE SANDBOX INCIDENT

I was born in 1948, to Paul and Isabel Taylor in North Tonawanda, New York, a small city situated between the larger cities of Buffalo and Niagara Falls. Virtually from birth I was expected to become an engineer- it was undoubtedly in my genes. Paul Taylor was an aeronautical engineer, a 1930s graduate of Parks Air College (now part of St Louis University). Paul had always been in love with aircraft. Before he went to college he had spent time as a 'barnstormer' and was copilot to an older pilot who flew an old WWI vintage Curtiss 'Jenny' trainer aircraft from town to town, selling rides in the aircraft to brave local residents. Prior to matriculating at Parks, Paul had worked as an auto mechanic and as part of his studies was required to obtain his license as an aircraft and powerplant mechanic. He had a burning desire to design airplanes, sell airplanes, work on airplanes, and do anything else involving airplanes. Although his father was a farmer who had emigrated from the British Isle of Wight in the late 1800s, the Taylor family in previous centuries had many members involved in the engineering field. The Isle of Wight is just a short distance away from Southampton, England, a bustling area of shipbuilding activity in times past. Ancestral Taylors had sailed with the British Navy and many worked at or near the Southampton yard on both commercial and naval vessels. In the back of one of Paul's workbench drawers there were some ancient pulleys and blocks, 'made by one of your ancestors a long time ago.' Research into this reveals a probable group of family members known as the 'Taylors of Southampton,' who were involved in the manufacturing and eventual mass production of ships' blocks for the British Navy in the 1700s.

DAMPERS: HOW VISCOUS DAMPERS PROTECT STRUCTURES

When Paul Taylor graduated from Parks in 1936 he went straight to work for Beech Aircraft in Wichita, Kansas, and in 1939 left for Buffalo, New York and a job at the Curtiss-Wright Company. The winds of war were blowing and Curtiss made military aircraft- lots of them.

When Paul and Isabel arrived in Buffalo they rented an upper apartment in the small city of North Tonawanda. The location was ideal, since Curtiss had six large plants in the Buffalo area and to reach any of them from North Tonawanda was relatively simple. If things didn't work out at Curtiss, there was always the other aircraft company in the Buffalo area, the Bell Aircraft Company, which also had several plants. During WWII more than 50000 people were employed directly making airplanes in the Buffalo aircraft plants.

Paul Taylor's dream of working on airplanes came to a sudden halt at war's end. The management at Curtiss had elected to maximize profits during the war by not introducing new aircraft. Instead, they stayed with their 'bread and butter' planes, the P-40 fighter and the C-46 Commando transports, which by 1945 were completely obsolete. Shortly after war's end Curtiss announced that it was closing its plants and was no longer in the business of making aircraft. Paul decided to stay in Buffalo and found himself a new job at the Wales-Strippit Company, a manufacturer of machine tool products that were used heavily by the aircraft industry. Paul thrived at Wales-Strippit, filed for numerous patents on behalf of the company and rapidly rose to the position of Vice President.

It was an application at Wales that piqued Paul's interest in fluid mechanics: liquid springs. He got Wales involved in the production of the liquid spring, a product that would become the first offering for Paul's own company, Taylor Devices, Inc.

The liquid spring in the late 1940s was little more than a laboratory curiosity. In the 1920s Sir George Dowty of England had discovered that he could make a very compact spring element by compressing fluids of the silicone family.

The big problem was that for maximum volumetric efficiency Dowty had designed liquid springs that ramped through a static pressure range up to 300mPa- a pressure level usually associated with the firing pressure of a military rifle.

Wales-Strippit's main product lines were punching and stamping equipment, and most of these used high-force coil springs to strip parts from their dies after metal forming was complete. In many applications, higher spring forces were required than could be easily obtained with coil springs. The only solution at that time was to make their products larger than necessary; just to hold a higher number of coil springs.

While looking for ways to obtain higher spring forces Paul discovered some old patents for liquid springs. After looking at the prior designs he decided to start with a clean piece of paper and design a low-cost liquid spring from the ground up, using modern materials and aircraft design techniques.

Paul's prototypes worked surprisingly well, with such high forces that one liquid spring could replace ten mechanical coil springs of the same overall size. Numerous patents were filed by Paul on all possible variations of this basic design, assigned to Wales-Strippit, and sales volumes soared.

I was born in 1948 and my sister Joyce in 1950. Our family had outgrown the apartment and in 1950 moved to a house that Paul had built on suburban Grand Island, about 10 miles from his job at Wales. In 1951 we received our first 'big kids' toy from Dad- a sandbox. This measured roughly 7 feet long by 4 feet wide, was made of thin-gage metal with a wood frame and was evidently something Paul must have discovered in the garbage heap at work. Nevertheless it promptly became our favorite toy and we were always inventing new games to play in our sandbox.

One day in 1953, when I was 5 years old, Joyce and I were playing in our sandbox when two cars entered our driveway, one was Dad's Plymouth station wagon and the other was quite shiny, new and occupied by four men. Dad and the men got out of the cars. All were wearing suits. They approached the sandbox. Dad came up to us and exclaimed, 'Hey kids, me and these other men want to borrow your sandbox for a bit; you can stand over by the house and watch us. These men are engineers like me and come all the way from California. They're from a big company called Hughes Aircraft, and they're working on a new rocket and they want to see this special liquid spring I made.'

All of this seemed fairly innocent to Joyce and me. After all, these big people just wanted to play in our sandbox too. After a good hour of preparation, the men finished by connecting numerous electrical wires to something they had buried in the sand in our sandbox. Dad came over to us and said those words all children hate, 'You'll have to go into the house now.' Then he added, 'My new liquid spring uses the power of explosives to make it work very fast- and we're going to test it now.' Joyce and I went into the house and watched the 'test' from the dining room window.

The men had a large battery that they carefully connected to some of the many wires coming out of the sandbox. Some of the men were watching some sort of box with many dials on it. When they were ready, all the men stepped back about 15 feet from the sandbox and Dad closed the switch to detonate the explosive. Unfortunately, he had used too large a charge- there was a loud 'whoop' followed by a 'crack,' and a small mushroom cloud of sand rose some 5 feet into the air. When everything calmed down Dad ran to the sandbox and began combing through the sand looking for the remains of the liquid spring. More important to Joyce and me was that one side of the sandbox was blown out entirely and piles of sand were now on the ground. The engineers from Hughes seemed totally unperturbed by this. One said 'Let's try this again with a much smaller charge, it really looks promising.'

This event served to introduce me to the field of engineering. I remember thinking two things:

1. Blowing up things is fun, so I should be an engineer.
2. Hughes Aircraft people are truly wonderful. You can make a mistake, but if they like your idea, they don't care. I'd like to work with them someday.

3. GROWING UP IN THE 1950s AND 1960s

Paul Taylor left his job at Wales-Strippit in 1955. The company was being taken over by a larger firm, who evidently did not look too fondly at Vice Presidents who were also inventors. In any event, Paul had already founded his own company, Taylor Devices, Inc., in 1954, which was formally incorporated in 1955. He rented space in an old lumber mill office building at 188 Main Street in North Tonawanda. This building had been rented previously by a printing company and heavy concrete pads had been poured to support several printing presses. These proved ideal foundations for the lathes and mills required to make liquid springs. Taylor Devices' first employees included one machinist, one sales person, one bookkeeper and Paul. As Paul went about inventing things, he brought vast accumulations of 'stuff' home that he stored in the garage. I was told that as long as I didn't hurt myself or break anything, virtually all the 'stuff' in the garage was fair game. And what a garage it was…in addition to a huge supply of hand tools, a complete suite of power tools, and box upon box of interesting items, Paul had a small machinist's lathe, drill presses and a grinding machine. While Dad was at work, I became proficient with all of the tools and learned as much as I could by disassembling and reassembling all the 'stuff.'

At age 11, I decided that I needed a boat, so with the aid of saved Christmas and birthday money, I bought a well-used 11-foot wood runabout for $45, and a worn-out 1938 Muncie Neptune 5 horse-power outboard motor for $15. A year later, using all of my mechanical ability, I managed to fix the leaks in the boat and got the motor to occasionally run intermittently on one of its two opposed cylinders. I finally asked Dad for assistance, but he claimed to have little knowledge of what was wrong, probably because he thought the motor was beyond help. Instead, he said 'Let's go over and see my friend Ed Ebert; he knows about outboard motors.' So Dad and I packed the motor into the car and went to see Ed at his home.

Ed Ebert was a patent draftsman by trade, and was also a self-taught engineer with numerous patents. Dad went back to work, leaving me with Ed, but before he would look at the motor, Ed insisted on taking me to his upstairs office where he did his work. I was amazed by watching Ed sketch with a pencil on the Bristol board medium of the patent office. This was an education in itself. He showed me how he finished his drawings by carefully inking over the pencil lines with a twin-tined India ink pen (the ink is held between two flat spring steel tines by capillary attraction, and a small screw and thumbwheel are used to adjust the distance between the tines to set line width). Ed let me try this and I quickly botched a line with an ink blob, ruining the sketch and the Bristol board. Ed asked me if I had thoughts about becoming an engineer. I said 'Maybe' and he exclaimed 'If you do, then always do what that sign on the wall says- read it, please!' The sign read:

Always be kind to a new idea...
...for you know not where it may lead you.

Ed said that the sign was given to him by his first boss, a Buffalo-based patent attorney. 'If I didn't keep looking at that sign, then I'd turn away many of the inventors who come to me wanting me to draw their inventions. It also works the other way; patent attorneys send me projects from inventors who hire them, and I have to figure out how to make their ideas into drawings.'

Ed and I went back to his basement, and within 3 hours and about $8 in parts the motor was running. I left Ed's house that day with two items: one outboard motor that was now running, and the beginning of a vision for my career.

Over the next few years I progressed through a series of boats and became a proficient mechanic. At age 16 my parents decided to nudge me a bit more toward engineering and I was given marching orders to go to work part-time at Taylor Devices.

4. 'ALMOST' AN ENGINEER

In mid 1964 I was on top of the world. I was 16, had my driver's license and a job. Even better, the job fully utilized my skills as a mechanic and the math capability that I was learning in high school, along with mechanical drawing and shop courses.

Dad assigned me to a Taylor Devices research group, where I was working for an eclectic group of engineers who worked on a myriad of strange assignments. Some of these were far ahead of their time. Here are some of the projects that were being worked on by the six-person research group in 1964:

1. A reciprocating internal combustion engine having no flywheel or crankshaft. Instead, a liquid spring with a tuned mass free piston would replace all of these parts.
2. A geothermal energy source using a liquid spring as its primary element.
3. An electronically controlled automotive suspension system using variable-rate liquid springs and variable fluid damping.
4. A mass-produced small liquid spring for manufacturing machinery that would be competitive with plain wire wound metal coil springs.

When I started with the research group the engineers briefed me on the four major projects. The project with the biggest immediate problem was the last one; the mass-produced small liquid spring. Dad had just put the unit into production and was having many problems. The engineers thought that my mechanical skills should help, since the problems seemed to be due to 'shop errors.' The problem was that many springs were being sold and the company was losing money on each one built due to high labor costs and quality control problems.

Taylor Devices had moved the production machinery out of its original rented building shortly after the money from its 1960 public stock offering was received. A deal was struck with the City of North Tonawanda for an 8-acre site on Tonawanda Island, a small island located just a few hundred feet offshore and occupied by several industries. Dad bought his 8 acres from the City for $25000 with the promise to build a new facility on the site. This was done by 1961. Unfortunately, Dad decided to 'help' the A&E firm he had contracted with for the design, thereby ending up about 300% over budget. To recoup the overage, his accountants insisted on very high overheads to pay the money off in a short period. The high overhead was causing the new commercial spring line to lose money with every sale. Demand was such that if prices were raised even slightly, sales dropped substantially. The project manager for the commercial spring line had attempted to do all he could to reduce overhead, and had even gone so far as to move the manufacturing out of the new building and back to the original rented building Taylor Devices had just vacated. This reduced the losses but still did not allow the product to turn a profit.

This would be my first real project with the company. After looking at the basic parts and arrangement drawings, I decided to talk with the assemblers. I carefully explained that I was not an engineer; rather, I was just a mechanic who wanted to make their jobs easier. This approach worked, and soon I was filling notebook pages with their list of problems. The biggest problem was that only one unit out of three would actually go together without having parts sent back for rework. The next major problem was that 50% of the units that eventually did go together were rejected for leakage. The solution to all this sounded relatively straightforward; the first step being to figure out why the parts didn't fit together.

My first step was simply to build the proverbial 'mechanic's dream'—a unit that was built entirely from perfect parts- or as close to perfect as could be found in the inventory. This unit assembled easily, didn't leak, and went directly into the cycle test machine. Not surprisingly, this 'perfect' unit worked well, and surpassed the required life cycle by a factor of two. Convinced that the basic design was sound, the engineers required that the test be repeated on all ten different sizes of the product. All ten units assembled and tested without difficulty when made this way.

The next step was to isolate which parts would prevent successful assembly and to test the unit when parts were made at the limit of production tolerances. The time involved to do all this would normally be very long, perhaps even years, except that I had told the workers that the problem was not them, it was the tolerances on some of the parts being too liberal. The production workers felt that I was a friend to them. They eagerly volunteered the specific parts that were troublesome, the dimensions that were causing the trouble, and whether the dimension gave trouble when it was on the 'highside' or the 'low side' of tolerance.

I verified their comments by measuring parts in the inventory with the suspected tolerance problems, and tried to duplicate the assembly issues. The product manager was kept informed of my findings, and came up with various ways to fix the parts. Some of these involved select fits, where specific high-tolerance/low-tolerance parts were match fitted. Other problems were addressed by simply reworking all the parts to tighten tolerances on the offending dimensions.

Within a 3-month period the commercial liquid springs were making a profit and I was ready to move onto another project. Dad's response was to give me a raise (a good thing) and tell the other engineers that I was not lazy like them and was not afraid to get my lily-white hands dirty (a bad thing). I was impressed enough by the success of this first project to add to my vision: *'Engineers should not be afraid to get their hands dirty.'*

DAMPERS: HOW VISCOUS DAMPERS PROTECT STRUCTURES

Like most people in a new job, I made the mistake of letting my initial success instill a large amount of overconfidence. Thus, for my next research project I offered to work on Paul Taylor's liquid spring internal combustion engine. This did not go as well as my first project, and eventually the entire concept was abandoned as it required too much money to complete.

Similarly, a review was also held on another of the four major research problems- the geothermal energy source powered by a liquid spring. This program was cancelled before anything was constructed, simply because the engineers' calculations did not support the project's viability.

During the summer of 1966 I was assigned to the largest of the Taylor Devices research projects: development of an active suspension system for automobiles. This was a long-term project which had enjoyed some success- which meant a customer was found who would fund the research. In fact, the company had two funding sources. One, the Walker Manufacturing Company, made automotive exhaust systems and wanted to market shock absorbers along with their mufflers. The second sponsor was NASA, which wanted to develop controllable isolation systems for their space programs, most specifically the Apollo program with its gigantic Saturn family of space launch vehicles.

With the Walker funding, two 1964 Pontiac Tempest station wagons were used as test mules, one with an essentially stock suspension, the other with liquid springs. The decision to pursue a full suspension rather than a simple replacement shock absorber had been made by the higher-ups at Walker. This also made the entire project very complex.

We used the simple aerospace industry axiom of 'Take the money, build it, try it out, then fix what breaks.' This allowed Taylor Devices to find every potential design problem all at once. Design issues were as follows:

1. High seal friction makes the seal life good, but gives a poor ride.

2. Liquid spring damper elements generate heat, so the vehicle rises higher off the ground when hot as the fluid expands, and sinks when it's cold.
3. Packaging the liquid spring damper into the place where the shock absorbers were originally located is nearly impossible.

Walker lost interest in the project rather quickly and in an attempt to keep the money flowing there search group had attempted to fix all of these items at once. In doing so, they caused new problems to show up. Walker cancelled the project. However, the project continued with NASA funding, since many of the items to be developed for NASA use on the Apollo program could be easily evaluated and tested on a test vehicle. NASA would benefit from the resulting improvements and the new technology could be used on the Apollo hardware. Taylor Devices' benefit in the commercialization of the NASA technology was not only allowed- it was strongly encouraged.

I was assigned to a five-person team that worked with a new rear engine 1965 Chevrolet Corvair test vehicle that was put through an impromptu conversion. As the mechanic I installed additional steel bracing to the chassis at multiple areas considered highly stressed. In the front of the vehicle, which in the Corvair included its trunk (or 'Frunk' in the 1960s vernacular), I laid down multiple layers of fiberglass cloth to increase chassis stiffness, and this fiberglass was impregnated with high-strength epoxy resin. The reason for all this was that Dad had decreed that he would use an unsupported MacPherson strut suspension in the front, with no upper A-frame as in the conventional American double-wishbone suspension (Figure 1). The idea was that this would reduce vehicle cost (for commercial use), plus evaluate 'off-axis static and dynamic loadings at the piston rod to seal interfaces' (for NASA use).

The design selected was completely modular and capable of accepting electronic controls. The front and rear struts were liquid spring dampers, with upper and lower pipe connections to allow various struts to be interconnected. By properly sizing the port restrictions and using cycling control valves it would be possible to provide semi-active roll and pitch control, plus controllable damping. In addition, additional plumbing connections allowed multiple 'air over oil' accumulators to be selectively engaged, thus controlling the vehicle's spring rate. The design approach was to split the group into a mechanical team and an electrical team. The electrical team was in charge of control valves and control logic. The mechanical team did everything else.

Figure 1. Unsupported MacPherson Strut Suspension in 1965 Corvair

I was tasked with doing the actual installation, and the little Corvair was up and running within 3 months. But there was a problem with the electronics. After 3 months of study the team had concluded that control must be done by a computer- and the computer necessary to do the job was physically larger than the entire Corvair. There were also no control algorithms that were simple enough to be solved in a short enough time for real-time control. Shortly after these facts became known the mechanical team was ordered to finish up the vehicle with mechanical controls. Thus, the little Corvair was suddenly modified with numerous high-pressure valves. Those that needed to be opened or closed periodically used electrical operation with a switch panel inside the vehicle, so that the 'co-driver' could make the required valve settings literally 'on the fly.' Other valves located under the vehicle frame controlled parameters that could be changed only between test drives.

Amazingly, the whole concept worked rather well, and the little Corvair underwent nearly continuous testing of various elements throughout the remainder of the 1960s and well into the 1970s. Important design elements that we developed included:

1. the basic Taylor Devices monolithic seal design that is still used today in nearly all of the company's products;
2. numerous damping schemes applicable to a myriad of commercial and aerospace/military applications;
3. controllable spring elements that were not only used on the Apollo program but are still used today by NASA on its most modern launch vehicles;
4. silicone fluid compounds and mixtures capable of working at extreme temperatures;
5. internal connection detailing suitable for extreme cycle life without fatigue failure, still used today in all Taylor Devices products.

The Corvair project was long term, and since the vehicle worked well the research team members moved onto other projects, coming back to the Corvair only when new concepts were to be incorporated. Eventually, other suspension test vehicles were fielded to evaluate other concepts (Figure 2). Those included were:

1. 1969 AMX- This two-seat sports car was outfitted with a replacement style shock absorber.
2. 1970 AMC Hornet- This little station wagon was outfitted with replacement type shock absorbers and a prototype liquid spring bumper system for crash protection.

3. 1971 Plymouth Barracuda- This 'muscle' car was outfitted with auxiliary rear suspension spring damper elements. The spring rate was electronically controllable with three-stage 'bang-bang' servo valves (three separate and discrete spring rates).

Figure 2. Taylor Devices' Test Vehicles, 1972

In a parallel effort, the company established a long-term relationship with local auto racing driver and car builder, Jim Hurtibise. Jim entered the USAC Indy car race circuit in 1960, but had his career altered by a crash with an ensuing fire during a race. Although terribly burned over more than 60% of his body, massive skin grafts and rehabilitation allowed Jim to continue racing. Jim was the most optimistic person I ever met- to the point of having skin grafts done on his hands with the fingers pre-bent so that he could still hold the steering wheel on his race car firmly. Always looking for the next new racing speed trick, Jim allowed Taylor Devices to test our liquid springs on his race cars, at speed, during qualifying sessions for the Indy 500 and other renowned USAC races. Figure 3 shows Jim's 'Tombstone Special' equipped with a liquid spring suspension strut in the rear of the car. Figure 4 shows a dual-purpose race/parade car built for a member of the Shriners. Shown with a long 'parade' nose, the car participated in the pre-race parades put on by the Shriners before a USAC race. With the nose changed to a short 'race' nose, the car competed on the Formula-Vee racing circuit. This car was equipped with very compact liquid spring suspension struts that I had designed on all four wheels.

In 1969 and 1970, Jim's 'Mallard' Indy car was equipped with a semi-active liquid spring suspension, operated manually while Jim drove. I designed a simple thumbwheel driven actuator to add or subtract fluid from the rear suspension struts. This was used to compensate the vehicle ride height for changes in fuel load and track conditions during a race.

Figure 3. Rear Liquid Spring Suspension Strut, Hurtibise Indy Car

Figure 4. Formula-Vee Race/Parade Car with Liquid Spring Suspension
Paul Taylor (left), Jim Hurtibise (in car), Doug Taylor (right)

DAMPERS: HOW VISCOUS DAMPERS PROTECT STRUCTURES

Due to liability concerns, we never let Jim leave the liquid springs in for the actual race. Occasionally this led to heated discussions with Jim, especially on the several occasions when his lap speeds went up appreciably from the addition of Taylor Devices' components.

The year 1966 also marked several milestones in my life. I officially decided to become an engineer, and I had a girlfriend. Her name was Sandi Forbes. In 1971 she became my wife, and we were happily married until her death in 2013. The year 1966 also marked my graduation from high school. I had good high school marks, and applied to Clarkson University, Rensselaer Polytechnic, Northwestern, the Rochester Institute of Technology, and the local State University of New York at Buffalo. I was accepted at all five schools, and ended up selecting Buffalo because of its proximity to home and work, the reduced costs of being a commuter student, and Sandi, but not necessarily in that order!

The year 1966 also marked my first chance to work on actual production projects rather than research. My first production project work was on hardware for the upcoming Apollo One launch of NASA. Taylor Devices produced two major products for Apollo, and the company's success led into similar products in later years for the Space Shuttle program, plus various unmanned launch vehicles. Taylor Devices' mission on Apollo was removal and retraction of the umbilical cables and service masts between the launch vehicle and its launch tower, or gantry (Figures 5 and 6). On Apollo, these included both electrical and fuel line connections, plus service and astronaut access/egress platforms. The 'star of the show' was the huge Taylor Devices damper/retraction elements with both semi-active and active controls (Figure 7). These were normally used as a wind damper. If winds became high, the damping was controlled to limit motion between the delicate launch vehicle and the immense, very rigid launch tower. Total damper stroke was plus or minus 6 feet. When it was time to launch, a command signal caused the damper to instantly convert to a tension actuator, which rapidly disconnected the umbilical cables from the vehicle. The swing arm to which the damper was mounted would then quickly swing clear of the launch vehicle. Terminal velocity on the tip on the swing arm was greater than 30 feet/s, and a set of 18-inch stroke Taylor Devices shock absorbers had to smoothly decelerate the swing arm into its retracted position so a mechanical locking latch could engage as a safety lock.

Needless to say, most of Taylor's production people were busy on the complex umbilical dampers/retraction element, leaving the company short handed on the swing arm shock absorbers. I was assigned to the assembly and test group for the Apollo shock absorbers.

Figure 5. Model of Original Apollo Launch Gantry, Circa 1965. Photo Courtesy NASA Figure 6. Artist's Rendition of Apollo Service Arm and Taylor Damper/Retraction Elements Photo Courtesy NASA

DAMPERS: HOW VISCOUS DAMPERS PROTECT STRUCTURES

Figure 6. Artist's Rendition of Apollo Service Arm and Taylor Damper/Retraction Elements Photo Courtesy NASA

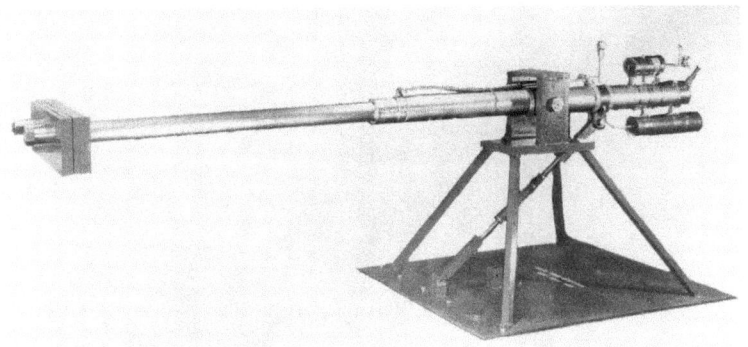

Figure 7. Apollo Damper/Retract Element, 1966

Testing the Apollo shock absorbers was an exciting proposition, using the company's new heavy-weight drop test tower. This test facility was funded by NASA and consisted of steel piles sunk until they contacted bedrock, a 50-ton concrete pad at ground level, and a 44-foot single steel I-beam tower, cable stayed in place. A weight of up to 6000lbs was cantilevered on small support wheels and winched up to the top of the tower. The shock absorber was instrumented and placed at the bottom of the rail (Figure 8). The weight was then released and would free fall up to 38 feet, at which point the shock absorber had to smoothly absorb its kinetic energy. After one got over the initial fright at seeing the drop weight falling from high in the air, the test itself was rather mundane- the Apollo shock absorbers had to work perfectly each and every time- and they did.

Although I thought this project rather insignificant at the time, it involved working with NASA and US government inspectors; my first real interaction with customers. All of my concepts of what was significant changed on January 27, 1967, when a flash fire occurred in the command module of Apollo One during pre-flight tests, killing astronauts Virgil 'Gus' Grissom, Edward White, and Roger Chaffee.

Figure 8. Taylor Devices' Apollo Drop Test Tower, 1966

Although this had nothing to do with the shock absorbers, or any Taylor Devices product, my mind rambled through all sorts of horrible scenarios, leading to nightmares and more than a few sleepless nights. The worst of these involved an umbilical arm swinging into a shock absorber I had built and tested, stopping the arm before it could latch, and the arm rebounding back out into the path of the rising vehicle- and its huge load of liquid fuel. To this day, I still often point out to workers at Taylor Devices the importance of building products that are as close to perfect as humanly possible. Simply said, 'Given Taylor Devices' customer base, if our product doesn't work—people may die.' This became a third item for my vision.

5. THE U/B YEARS

The 1960s offered wonderful opportunities to students from diverse backgrounds who came to New York State. For a commuter student like me, the cost of a university education was nearly zero if you had good marks in high school and you scored well on what was called the 'New York State Regents Scholarship Exam.' This exam was not particularly difficult for a good student. For example, in my high school graduating class nearly 40% of the graduates, including me, scored high enough to be awarded a Regents Scholarship.

In my case, after reviewing my parent's income tax returns, my Regents Scholarship was $350 per semester for 4 years. This may sound like a small amount, except for the fact that this was the 1960s(when gasoline was around 32¢ per gallon), and the State of New York subsidized its State University Centers heavily; so heavily that full-time student tuition at U/B was only $400 per semester. So in my case, as a commuter from home, the cost of my yearly university education was $100 per year plus food and gas for my car.

Because of the low-cost tuition, the University at Buffalo had a huge number of applications, and selection was made essentially on academic test scores. The end result was a wild and crazy student body of high intelligence with radical ideas…and I loved it!

After sweating through the first semester of my freshman year and discovering the level of work that was required, I found university life truly wonderful. To maintain a reasonable income, I arranged to work at Taylor Devices during the 4-week winter break and all summer. I structured my curriculum along fairly well-defined guidelines:

1. I was an Engineering undergraduate.
2. I enrolled in a Mechanical Engineering curriculum.
3. Major emphasis was Design Engineering.
4. Minor emphasis was Vibration Mechanics and Metallurgy.

This solidified my course requirements for what ended up to be a 5-year degree program for a BSME Degree. Although typical engineering course loads of today require roughly 128 credit hours for a degree, the program I had selected required a minimum of 144 credit hours, and I eventually ended up at 156 at graduation.

To me, school was still secondary in my life because I liked working at Taylor Devices much better. The company had decided that I got along well with customers, and that I could be assigned to summer projects involving a lot of fieldwork. Later, when the Vietnam War was literally tearing the country apart, I was one of a very small group of young engineers who would actually work on military projects.

Throughout the war, male students always had the military draft hanging over their heads—forcing them to either excel at school or be drafted. The rules were simple:

1. If you passed less then 12 credit hours of courses in a semester- you were subject to the draft.
2. If you flunked out or took a sabbatical- you were subject to the draft.
3. As the war raged on, graduate student deferments were cancelled- so if you graduated you were subject to the draft.

As would be expected, each male student who was in good health had definite incentives to excel at the university. The penalty for failure could literally lead to one's death in a far off place.

As for me, I was well aware of a type of draft deferment, authorized by the Defense Department, for those working on government contracts deemed 'essential to the war effort.' I decided to hedge my bet, so to speak, by being both a good student and a person who worked on essential government contracts. At Taylor Devices this was easy; the company made products for shock and vibration control and, as new weapons were fielded, shock and vibration problems were likely to show up. I spent the summer of 1967 working on the new CH-54 Sikorsky Skycrane Helicopter for the US Army.

Early on during the war, the Army decided it needed a heavy lift helicopter, and it needed it quickly. The helicopter manufacturers eagerly waited for an official RFP to be released for development of the new craft, but Igor Sikorsky, founder of the Sikorsky Aircraft Company, suddenly unveiled a new helicopter, already flying, with full heavy lift capability. Rumor had it that Mr. Sikorsky and some of his older workers built this aircraft during a 2-week summer vacation plant shutdown. As the story goes, they cut apart two helicopters on the production line and built the first Skycrane with the parts. In any event, the Army bought the aircraft and it went into production with little time for development testing (Figure 9). Service reports from Vietnam revealed a severe problem with the airframe, caused by a resonant mode excited by certain combinations of load and cable length. The resonance, accompanied by heavy shaking of the airframe, caused structural failure after only 40- 50 hours of flight time. Sikorsky's engineers decided that the best solution was to incorporate an isolator between the lifting cable and the airframe, within the mechanism of the cargo hoist winch. Required static force was in the 182kN range, with a 75mm stroke, and a so-called 'smart' damper. Under 3Hz resonant excitation and higher frequencies no damping was desired. Under shock loadings, moderate damping was needed. In the event of a dropped load, extremely high damping was needed to avoid winch failure.

Figure 9. Sikorksy CH-54 Skycrane. Photo Courtesy Sikorsky Aircraft Company

Figure 10. Sikorsky Skycrane Isolator, 1967

Sikorsky was aware of the work on active isolators being done by Taylor Devices, and requested a design solution. The proposed liquid spring isolator utilized a mechanical logic system to select zero, moderate, or high damping. Sikorsky then added additional requirements for keeping the isolator at the center of its stroke even though the load varied radically. This required an active actuator that could add or remove fluid from the pressure chamber, plus a sensor that would provide a signal to the control system that varied with the lifted load (Figure 10).

All of this was packaged into a compact but highly stressed device made almost entirely from heat-treated titanium. The device worked well and was soon in production. The only problem was that the device generated about

900N seal friction on a 182kN load, which still caused some shaking to be transmitted. Desired friction was zero, but this would cause seal leakage. The end result was that the Army wanted an improved isolator with less friction. At the same time, they wanted the load capacity to be increased by 20% with no change in package size or weight. The solution was a completely new design, and a radical one. The design hinged upon a piston rod of only a scant 12·5mm diameter to support a carried load up to 220kN that was shaking up and down continuously. The piston rod, and much of the balance of the unit, was constructed of a new 'super' material known as 18% nickel maraging steel, rated at a yield strength of 2000Mpa. Prototypes were built, and two major problems arose. One was that seal friction was still too high, at 450N. The other was that the slender piston rod kept failing where it attached to the load clevis.

My task was to fix those problems- and quickly. Using the knowledge gained from all of those experiments on the Corvair suspension project I was able to alter the seal design and material to get seal friction down to a mere 75N, well under the 180N limit imposed by Sikorsky.

Fixing the piston rod proved much harder. The initial design used a male thread turned on the end of the piston rod with a small enough diameter so that the sharp threads would clear the soft pure Teflon® seal as it was pressed onto the piston rod. The thread had a properly designed thread under-cut, and this was the failure point. At Sikorsky's request we experimented with a shear weld with electron beam welding for the connection, but the required weld depth was so great that the electron beam could not be consistently aimed at the desired interface. So, in desperation I tried what is termed a 'bastard' thread—one that follows no standards and is specific to a single application. My thread was actually larger in basic diameter than the piston rod; the trick was to truncate the thread by turning down its outside diameter to the pitch line diameter, the presumed center of load for the engaging threads. In view of the expected fatigue loadings, this truncated thread was mechanically rolled, rather than cut, and had no undercut. Rather, it was smoothly blended by tapering it into the base diameter of the piston rod. To assemble the seal onto the rod I left enough clearance to put exactly one wrap of painter's masking tape over the thread. The seal would slip over the tape, provided you heated the seal in hot water for 10 minutes before assembling. This redesigned piston rod worked well. In fact many of these Skycrane helicopters are still in service today and are used in construction and fire fighting.

When the unit went into production the Army asked for one last test: a gunfire test. They were concerned that high-strength steel was brittle and would explode if hit by gunfire. The test was to load the liquid spring up to its proof pressure, which was 350 MPa, and hit it with 30 rounds of 7·62 mm NATO standard rifle ammunition. We tested at an open area near Sikorsky's plant, with a Connecticut state trooper as the shooter. Ed Dudek, the Sikorsky staff engineer and I crouched behind a concrete barrier to watch the test. At the conclusion of the firing, the pressure gage attached to the unit still read 350MPa. Ed said to me, 'Okay, it passed. Now you go out and pick up the unit and put in the truck so we can take it back to the lab.' I looked at the unit- and then looked at Ed and replied 'Ed, you get to pick up the unit- for two reasons. One is that I am a lot younger than you and have a lot of life left to live. Reason number two is that I read your contract with Taylor Devices on the flight from Buffalo and this unit became Sikorsky property when you received it.'

The actual test results are still classified; the unit was taken by Army personnel to an undisclosed lab; so I can't comment further on the test results. However, we did settle the immediate problem of relieving the 350MPa pressure by doing the obvious: we had the state trooper shoot the test pressure gage off. A cloud of fluid vapor surrounded the unit as it rapidly depressurized, making it safe for transport.

The aerospace industry is very much people oriented, and we became known throughout the industry. The following summer of 1968 had me working with Sikorsky's competitor, Bell Helicopters out of Fort Worth, Texas, on the

Bell 'Huey,' the workhorse helicopter for carrying troops in and out of battle during Vietnam. The problems on the Huey related to its crashworthiness when it lost power due to battle damage. Secondary issues related to adding some armor plate to protect the pilot and some other items against the Russian AK-47 rifle round. The enemy had found a weakness in the Huey. They would shoot upward and place a burst from their AK-47's into the bottom of the aircraft under the crew. If they could hit the pilot's seat pan, the round would penetrate the pan and potentially hit the pilot's buttocks. Because the AK-47 round becomes quite unstable when entering flesh, the bullet yaws and tends to move around until it finds a solid object- like the pilot's spine. This problem was fixed with additional armor plate. The other armoring and crashworthiness problems were also easy to fix; however, the results of this project are still classified.

The summer of 1969 marked my first project with the US Navy, and the beginning of a successful 35-plus year working relationship. Our first project was to modify three high-speed boats (similar to WWII 'PT' boats) to operate at 80-plus km/h speeds in the open ocean while carrying a full complement of torpedoes. The inputs to the hull under high sea states were so severe that the ship deck sections carrying people needed to be base isolated to prevent crew injuries. Taylor Devices provided an isolation system consisting of multiple semi-active hydro-pneumatic suspension elements and anti-roll mechanisms, using multiple steel levers or links. My task was to help the Navy engineers assemble and rig the numerous suspension items, and also debug and tune the system during sea trials. Our work area was far to one side of the old General Dynamics Quincy Shipyard in Massachusetts, cordoned off from the public and the shipyard workers. We completed one boat. The sea trials were excellent but our project was suddenly cancelled by our sponsor. The boats vanished, and Taylor Devices was ordered to shred all drawings and files on this project.

I had a second project in 1969, related to Taylor Device's industrial products. Eventually, this led to my first US patent, the first of the 32 patents I hold today. I had completed my courses at U/B in fluid mechanics, and had learned many interesting things about high-velocity compressible fluid flows. One of the oddities of engineering is that systems theory likes to use so-called viscous damping, where force is proportional to velocity. What makes this an oddity is that hydraulic dampers did not normally provide this type of output since it is not cost-effective. This is because hydraulic dampers must operate at very low pressures to provide linear viscous output, requiring huge dampers to provide little force. Indeed, the upper limit of useful viscous flows is about 1500kPa. Above this value, flows are largely inertial, where damper force is proportional to velocity squared. This type of 'V-squared' hydraulic damper is not particularly effective for real-world applications, since in any given reciprocating cycle it absorbs fairly small amounts of energy compared to a linear 'systems style' damping element. A really effective damper normally requires an exponent of 1·0 or less, such that when subjected to cyclic loading it dissipates a large amount of energy without high transmitted forces. Conversely, if one looks at a hysteretic damper in comparison, dissipated energy is very high but transmitted force remains high even at tiny input levels. Indeed, typical automotive dampers after decades of evolution usually have a damping exponent of about 0·6 irrespective of manufacturer and vehicle. In an auto damper this is achieved by using mechanical valves, usually of the spring-loaded disk or poppet type, working at pressures in the 3500kPa range. Conversely, standard Taylor Devices' products optimize at much higher damping pressures in the 40000- 140000kPa range- beyond the capacity of conventional valves. The product line I was assigned to was called 'crane buffers,' a large industrial shock absorber used to protect steel mill buildings and large overhead traveling bridge cranes from crash damage (Figure 11). Output forces for this product range from 15 tons to nearly 300 tons. The Taylor Devices crane buffer was fairly traditional. It used an internal 'piccolo tube'- an internal pressure tube with orifice holes drilled in a spaced pattern. When impacted, the piston head pushed fluid through all the holes, but as it travels down its bore it progressively pushes fluid through fewer holes. Properly set up, this device can achieve equivalent damping exponents in the range of 0·3- 0·6,as long as the input motions are fairly well known. Since each discrete orifice hole has Vexp2 flow, if you stroke this product at 2·0 design speed its force would go up by 2·0 squared, or 4. Since a manufacturer could not afford to put high safety margins into a competitive item, it was critical to know what the customer's input was. And, in the steel mills, no one ever could be sure at what speed a crash could occur.

I borrowed heavily from my U/B course information and conceived a converging– diverging duct type orifice designed to operate at fluid speeds above Mach 1 (the speed of sound in the fluid). For the silicone-based fluids used by Taylor Devices, Mach 1 occurs at around 50000kPa pressure differential. The trick is how to handle the shock wave that forms inside the device on the exit side of the orifice as the flow goes into the sonic range. I developed a design that would allow the orifice to go smoothly through Mach 1, at which point many interesting and useful things occur. My calculations predicted that a single properly sized orifice of this type should produce damping exponents a slow as 0·3 if operated above Mach 1 fluid speed.

Figure 11. Taylor Devices Crane Buffer arresting the motion of a hot coke container

The first tests on this product were performed by crashing a full-size railroad flat car into a hard stop at 10mph. This was followed by testing the buffer between two railroad engines impacting head on (Figure 12). It's a long story of how all this happened for a first test, but essentially the new orifice was installed into an experimental Taylor Devices product that wasn't working very well, with testing funded by a very large sponsor. Tests with the new orifice went perfectly, and flow exponents in the 0·3- 0·4 range were achieved. Feeling that I had total knowledge of how the new orifice would behave, the second group of tests were made on a modified Taylor Devices crane buffer, tested on our drop tower.

We tested multiple impacts to an equivalent fluid flow speed of Mach 1·2 with no problems and near-perfect results. Being overconfident, I decided to skip additional data points and raise the drop weight high enough to push flow speeds to Mach 3. Unfortunately, it appeared that the shock wave forming downstream of the orifice didn't form in the right place at Mach 3, and the unit reverted to V-squared flow, causing a total failure of the unit, the load cell, and the test fixture. Output force measured was seven times greater than expected. As a lesson to myself, I saved the bent-up piston rod from this test unit, and it resides to this day behind my desk at the office. It's there to remind me to be humble!

My father became intrigued with the new orifice design- at first he called it 'a fixed orifice that isn't.' This transitioned to 'fluid amplified orifice,' which eventually became simply a 'fluidic orifice,' as it is known today. Given my basic concept, he promptly conceived an improvement in the design, causing me in turn to improve on his concept. Eventually we took all of the concepts to the corporate patent attorney for filing as a series of patents that were issued throughout the 1970s.

Figure 12. Testing a Crane Buffer with Two Locomotives

During 1970 the company was contacted within a 2-week period by all four of the Detroit automakers to develop an energy-absorbing 5mph bumper for cars. The word was out that new federal regulations were being drafted due to high auto repair costs from low-impact-speed crashes.

Taylor Devices produced bumper shock absorber prototypes for all four manufacturers, which underwent extensive secret testing throughout 1970 and 1971. At the same time, the automakers undertook an extensive lobbying campaign to undermine the proposed federal legislation.

All the tests went well, using two relatively small liquid spring damper elements on each bumper, which also served as the attachment for the bumper to the vehicle. The shock absorber was essentially a small crane buffer with a liquid spring reset combined with the new fluidic orifice. Roughly 50mm in diameter and 250mm in length, output force was 5 tons over an 85mm stroke. The successful tests results were kept under wraps by the automakers, and Taylor Devices' contracts with them prevented public disclosure of any type. Somehow, information on the tests was eventually leaked to the

Insurance Institute for Highway Safety, probably by a test engineer from one of the automakers. In due course, the Institute leaked their information to the US Senate Commerce Committee, who then subpoenaed Dad and me to appear before the Committee. In addition, Taylor Devices had constructed a bumper system and had it installed on one of our test cars (Figure 13); they subpoenaed this vehicle also.

Figure 13. Hornet Bumper Test Vehicle

The Commerce Committee turned our testimony into a festive occasion, culminating in running a demonstration for the press. The senators accompanied Dad and me to the parking area outside the old Senate Office Building in Washington. Senator Phillip Hart jumped into the car in the passenger's seat; Dad drove and rammed the AMC Hornet into the side of the building. I was busy with the press giving them facts and commentary. The test went well- no damage to the car, the building, or the Senator (Dad figured he was expendable for the publicity value). That night the crash demonstration made the national news on all three TV networks, and virtually every major US newspaper had our trusty Hornet on the front page the next morning.

The end result was that the legislation passed; but three of the four automakers now hated Taylor Devices. The fourth, little American Motors, eventually saw a jump in Hornet sales from all this publicity, and remained friendly; the end result was a nice long-term contract to Taylor Devices for the majority of AMC vehicle bumper shock absorbers for model years 1972 through 1976. Eventually the federal legislation was 'watered down' in 1976 after intense lobbying by the automakers, but by that time more than one million bumper shock absorbers had been built by Taylor Devices in joint ventures with Rockwell International and Consumer Glass Limited.

Shortly after the bumper shocks hit the press I began work on a related publication for the Society of Automotive Engineers. Professor Nelson Isada at U/B, my instructor in shock and vibration, somehow found funding to help me perform crash tests at U/B (Figure 14). As a result, a 2-ton crash test sled riding an inclined ramp was constructed within the Parker Hall lab at the university. I teamed up with another student and we ran numerous crash tests into spring and damping elements with a very rudimentary 'first-generation' human test dummy (as an aside, this dummy, nicknamed 'Fred,' still can be seen today on the crane runways at the U/B Earthquake Test Lab). The resulting paper was accepted and published by the SAE in late 1971 as my first publication.

6. ENTERING THE CORPORATE WORLD

In 1971 I began work at Taylor Devices full time, and promptly married Sandi, who had put up with little more than an engagement ring the entire time I was working on my degree at U/B. Receiving the BSME degree was anticlimactic; I had been working as an engineer for several years, was published, and had patents pending. Although I was approached by several universities offering graduate programs, I had no real desire to continue my studies. I looked at what courses were offered at the graduate level in shock and vibration and was unimpressed. I felt I could learn much more in the field.

At this time the company was in a transition, and a difficult one. The company managers had decided that the Vietnam War would last many more years, and had postured the company towards military contracts. When I started full time, layoffs were starting since the military work had literally vanished as the war began to wind down.

I expected to start in the engineering department, but instead found my Dad in one of his truculent moods. He called me into his office and said, 'Sure, you got an engineering job here, just pick which existing engineer you want laid off and you can have his job.' I must have had a sick look on my face, since the engineers were my friends and colleagues, but good old 'P.T.,' as he was called at work, had more to add. He said, 'If you want a job in the sales department you can have it, and I won't lay anyone off since we fired all our salesmen last year to save money!' Thus I began my corporate career as a salesman for a company that had no sales department in a period when their major market was shrinking.

Figure 14. Doug Taylor Crash Sled Testing at U/B, 1971

Other than the custom aerospace products, Taylor Devices' only other market was its steel mill crane buffer line. Fortunately, worker safety was becoming a major issue in the USA and rumors were that new safety regulations were being drawn up called the Occupational, Health and Safety Administration Requirements, or 'OSHA Regulations.' As luck would have it, prior long-term success of Taylor Devices' crane buffers had so much impressed some of the steel mills and the spec writers that the new OSHA regulations specified that all continuous-duty cranes required energy-absorbing bumper systems. Taylor Devices had several competitors in this end of the business, all comparably priced, but we had a big surprise in store for them. I sat down with the engineers and within 3 months we had redesigned all of our crane buffers to incorporate our most advanced low-exponent fluidic orifice. The cost savings were almost 50%. As a result, just before the OSHA handbooks were issued, Taylor Devices introduced a full line of crane buffers priced 30% under the competition (I figured we could keep the other 20% for ourselves).

DAMPERS: HOW VISCOUS DAMPERS PROTECT STRUCTURES

Crane buffer sales soared and became a solid 75% of Taylor Devices' business. Our industry market share eventually reached an astounding 90% due to our established high quality, proven reliability, and the reduced price.

With the business now improving, I began looking for new markets and found two: the US Navy and Air Force. I convinced the company accountants that they would see higher sales if I brought in another engineer just to concentrate on crane buffers, and they agreed. As steel mill sales expanded throughout the 1970s, several more sales engineers were added.

In 1971 I was contacted by General Dynamics Pomona Division in California and asked to develop an isolation system for a project known as the Phalanx Gun. This was a ship's close-in defensive system and consisted of a modular high RPM 20mm Gatling style gun (having multiple revolving barrels) with closed loop radar fire control (Figure 15). The gun would be used as a 'last-ditch' defense against incoming missiles and attacking aircraft. The Navy had insisted that the weapon be subjected to shock survivability tests in accord with Military Standard MIL-S-901C. This required putting the entire weapon on a small barge, in fully operational form, and subjecting it to multiple underwater detonations- the closest being a charge equivalent to roughly 90lbs of TNT detonated at a range of 20 feet from the barge (Figure 16). Previous tests on other systems revealed an expected peak translational velocity in the range of 3m/s, with roughly 50 total cycles of motion. Other design issues were that the isolation system must not stroke under wave inputs in the 3 g range, along with the usual list of severe military environments.

This was the 'slide rule age,' so no formal computer analysis was possible. I considered all the available test data and then designed a hybrid isolator. This consisted of a hydro-pneumatic spring preloaded to an extended position via an internal hard stop, combined with an internal parallel low-exponent fluidic damper. The isolators were installed vertically at the four corners of the system, and connected to the ship's deck through a support structure. At the top of the isolator was a series-mounted three-axis rubber isolator that would allow some isolation in the horizontal plane. Since the detonation forced primarily vertical motion of the barge, the rubber isolators were small and the hydro-pneumatic isolator was large. Proper operation of this device was based on the assumption that for an underwater detonation the gas bubble that is formed finds it easiest to initially displace water upwards.

Figure 15. Phalanx Gun. Photo Courtesy General Dynamics Corporation

Figure 16. Navy Livefire Shock Test, Test Object Under Security Screen

DAMPERS: HOW VISCOUS DAMPERS PROTECT STRUCTURES

I was uncertain as to whether the California aerospace industry and the Navy would accept a concept like this. Thus, I contacted Taylor Devices' West Coast representative, Dr David Lee, to help me sell the system. I told David that I could handle the Navy and he could work with General Dynamics directly.

The Phalanx was completed and was shock tested near San Clemente Island, off the California coast. The tests proved highly successful. The Phalanx incurred no damage and was fully operational after the test. Naval personnel told me that they never thought a complex system such as this would ever pass MIL-S-901C tests, and indeed, the Phalanx was the first modern electronically controlled system to ever pass the test and remain operational. The Phalanx went on to become very successful, with nearly 600 of them installed on the warships of several countries. Virtually all US naval surface combatants are equipped with one or more Phalanx systems.

David Lee and I found that the success of the Phalanx generated numerous inquiries from other aerospace companies building shipboard components. Most of these led to new projects. In later years, virtually every electronics cabinet put on a US naval vessel included either complete isolators or dampers from Taylor Devices.

One of these projects was related to so-called 'breakaway' isolators, where the isolated system remains completely rigid relative to the ship's deck until a high-level shock occurs. The most elegant of these is the so-called Hexapod Isolation System, providing six-axis shock protection along with rigid pre- and post-shock positioning to a mid-stroke position. Patents were eventually issued to both David Lee and myself on this isolation system. Typically, a Hexapod uses six hard-centering liquid spring damper isolators. Centering is obtained via a mechanical mechanism, so precise that on ship's navigation systems the pre and post-shock alignment is held within 0·002mm axially and 0·003 degrees in angular position relative to the location of the ship's keel.

While working on the initial Phalanx designs, I received an odd phone call from some people at the Wright–Patterson Air Force Base in Ohio. They invited me to come to a so-called 'think tank' which would study and evaluate basing modes and configurations for nuclear ballistic missiles. This seemed like it could lead to some business so I agreed to attend. The group consisted of approximately 30 people representing numerous aspects of the US nuclear weapons programs. I was in a small subset of this group evaluating potential isolation systems. From the Phalanx, I was fairly familiar with the explosive motions used for shipboard inputs, but after seeing what near-miss nuclear weapons ground motions looked like I was uncertain if it was even possible to protect against these inputs.

Without getting into restricted material, imagine that you are sitting in a chair in your office and the ground suddenly drops away more than one meter, almost instantly, at an average speed of 10m/s. Then the motion suddenly reverses and the ground comes back up 1·2m, at an average speed of 7m/s. This is the first part of a typical nuclear weapons ground motion, the first part of what can be several cycles of motion. A nuclear missile and its launching system must survive multiple attacks from various positions before its ground crew receives an order to launch (remember that US doctrine during the Cold War was not to launch a nuclear strike until an enemy had detonated a nuclear weapon over or on US soil). The various detonation positions were three-dimensional, since a warhead could be set to detonate after ground penetration, at ground level, or at high altitudes. Compounding the problem was that an enemy could be expected to attack any single US missile with multiple warheads, from multiple positions, either simultaneously or in a timed sequence.

At the time this project began, the front-line US land-based ballistic missile was the Minuteman III. Our proposed missile for the think tank project was called the Minuteman IV, and it did not necessarily use a fixed launch base; rather, it was a fully mobile missile and launcher that could launch either from a formal base or from open terrain.

The requirements given to us were to design a suspension system for this vehicle that would provide a reasonably good ride under off-highway road conditions, plus be capable of allowing the missile to survive a full-blown nuclear attack without damage. This was a multi-year project. Eventually the design was completed and isolators were built. These were roughly 0·3m in diameter with a stroke of 1m, and an output of 1800kN (Figure 17). At the time this was just another one-off custom design by Taylor Devices, but this particular isolator would indeed be very significant in later years. To maintain secrecy, procurement was made on a sole-sourced basis with no item description. In this type of procurement, supplier down-select is made based on previous experience, demonstrated technical expertise and quality assurance capability. This down-select was evidently made before I was even invited to join the think tank. At that time, Taylor Devices was already well known for its technical and manufacturing capability. The company's quality assurance programs were in compliance with US military and NASA quality standards. When it came time to procure, I negotiated a rough price with the Air Force officer and we agreed on an item description- literally something like a 'number 4186 widget.' A buyer would then call Taylor Devices and ask for a price on that item. The previously agreed-upon price would be quoted and the buyer would state that the price was acceptable. We would then receive a purchase order, not necessarily from the Air Force, but potentially from a designated third-party firm.

Figure 17. Minuteman IV Missile Launcher Isolator

From the viewpoint of shock isolation, the Minuteman IV was perfectly viable, but for rumored political reasons it never went into full-scale production. The program slowly faded from view, even though the isolators were successfully tested by the Air Force using multiple inputs at full weapons-grade velocities. All in all, the new isolator seemed virtually bulletproof and proved capable of attenuating virtually any conceived input—we just did not have a customer for the production units. As for me, I simply moved onto the next project: I had received a call from NASA and was asked to join a design group of NASA suppliers at Cape Kennedy for a new program, the Space Transportation System (STS), known today as the Space Shuttle.

7. THE SPACE SHUTTLE

My first meeting at 'The Cape' on the Space Shuttle proved surprising. The base security levels surpassed that of both the Air Force and Navy bases I had previously worked at. Much later the world found out that this was to keep the program information away from the Russians, who eventually announced their competing 'Buran' program, which was never completed.

When I arrived at the assigned building, my pocket wore no less than four security badges, sequentially placed in a specific order. An information officer addressed our group of five engineers, representing NASA, Rockwell International, Planning Research Corporation, Taylor Devices, and the University of Wisconsin. Our task was the design of the tail service modules (TSM) located on the launch platform and interfacing with the tail of the orbiter; right next to the giant exhaust nozzles of the Shuttle's liquid fuel rocket engines. The TSMs were part of the Shuttle's MLG (mobile launch gantry). In reality, the MLG is a huge, oddly shaped building (Figure 18). Underneath the platform that supports the Shuttle are numerous rooms and equipment areas. The Shuttle itself is an immense vehicle and a true 'space ship' in the classical sense. Quoting astronaut and U.S. Navy Captain William Shepherd, 'Imagine a two-million kg vessel, the size of a World War II vintage destroyer but turned vertical. Then realize that the Shuttle is so powerful that it accelerates from zero to 1200km/hour in a mere 40 seconds.'

From the mock-ups we developed, the TSMs were essentially small steel buildings with domed roofs. Inside the TSMs would be numerous dampers and shock absorbers. Virtually at the moment of engine ignition large gravity weights inside the modules would free fall, pulling on a slack cable attached through a rotating mast to the umbilical arms protruding through a rotating open dome at the top of the TSM. This dome formed a section of its 'roof.' At nearly the same release time as the gravity weights the rotating dome section would also be released, and it would begin a free fall rotation towards the umbilicals.

Figure 18. Space Shuttle and MLG

When the slack was out of the cables attached to the falling gravity weights the impact to the umbilical would actuate disconnect latches on the Shuttle, and the gravity weight pulling on the mast with the attached umbilical arms would pull the umbilicals inside the TSM, just as the falling domed roof section, called the bonnet, impacted against two 'bonnet buffers,' latching into a closed position and sealing the TSM against rocket blast.

Meanwhile, inside the TSM the falling gravity weight hit a 'counterweight buffer' and the mast would ram into two 'mast buffers.' Later on we discovered the disconnect shock loads on the umbilical latches were excessive, so we added 'latch shock absorbers.' We also had a timing issue. We were allocated less than 1 second to accomplish all of the noted operations in sequence. If the bonnet failed to close, or the umbilicals did not get retracted, total loss of the vehicle could occur.

Although this sounds complex, in reality it was the only system we could conceive that could be designed with the 100% redundancy that NASA demanded. In essence, the TSM mechanisms are similar to a gigantic mechanical clock works, and when driven by gravity can be made extremely reliable.

We completed the design during the bicentennial year 1976. This was the year the large American flag was painted on the famed NASA Vehicle Assembly Building (VAB) dating back to the Apollo program. This building would also house the Shuttle. One of our favorite pastimes was to take the elevators and stairs to the roof of the VAB and have lunch in the sun while leaning over the edge of the roof to talk to the two guys who were the winning bidders to paint the flag on the building's side.

During the course of the work on the TSMs, other groups often asked my opinion on shock problems. As a result, the upper oxygen vent arm at the top of the Shuttle's main fuel tank eventually incorporated long-stroke dampers to ease the huge vent arm into its stowage latches under all conditions of launch-day wind loads.

Later, towards the end of the design effort, I was approached by a group designing swinging access arms for crew ingress/egress and maintenance. Because the entire program was overrunning its design funds, the group needed shock absorbers at almost no cost. After pondering this problem, I asked if NASA had an area where they stored leftover hardware from Apollo. NASA personnel answered in the affirmative and so we took a short expedition to an outdoor storage area. All of the Apollo shock absorbers and dampers were exposed to the blast of the Saturn rocket during launch, but had finished their duties by the time the spacecraft began to move off the launch pad. Because the units were subsequently 'cooked" by the Saturn's exhaust blast, they were used for only one launch. What I had remembered from my days at Taylor Devices on the Apollo program was that each launch required one full set of shock absorbers for the launch, a second full set for pad spares, and a third full set for depot spares. As luck would have it, we located a good supply of old Taylor units and they appeared ready to use again.

When the Shuttle made its first launch, if one looked closely at the launch tower, completely painted gray, you could see our Apollo shock absorbers, still painted 'Apollo orange,' making a nice contrast. Eventually NASA assigned Shuttle part numbers to them, and the venerable Apollo units were repainted 'Shuttle gray.'

The TSM and upper oxygen vent line shock absorbers were manufactured by Taylor Devices in the late 1970s and early 1980s. The units have proven trouble-free and will last through the end of the program, as will the 1960s vintage Apollo shock absorbers carried over to the access arms.

8. THE MX MISSILE

When the design effort on the Shuttle program was complete, I thought I could relax a bit. The company was doing well and I had been promoted to Vice President. Those thoughts ended when I received a call from Martin Marietta Aerospace in Denver, Colorado. The caller explained that he was referred to me by a 'mutual friend' at Wright– Patterson, and that he had a design team that wanted to talk to me. I gathered Taylor's design engineers and filled them in on the old Minuteman IV project. We met with the Martin Marietta people and discovered that a new missile was in the works, called the MX (for Missile Experimental) and it had the full backing of President Jimmy Carter.

The MX missile is a technological marvel, and the initial basing mode selected by the Defense Department was from a fixed horizontal launch building. The Carter basing plan used 400 missiles scattered amongst 4000 launch buildings, all identical and connected by roads. The missiles could be shuffled from one launcher to another, making this the ultimate 'shell game.' The survival concept was simply that no enemy could afford to build all the missiles and warheads to positively destroy all the launchers. The problem was that there was political opposition to the program due to its immense cost. It might not be possible to build all the launch buildings and related infrastructure. Thus, the launch buildings needed an isolation system to protect the missiles against nuclear attack, even though the launch building's inside diameter was previously sized and the design frozen, allowing for limited 'isolator rattle space.' The dimensions were fixed and nonnegotiable, since that drove the entire design of the launch building. The missile diameter was also fixed previously and the actual missile was well into the design stage.

There is a time-honored set of three axioms for designing any shock isolation system, and these are:

1. Know the input.
2. Bound the output.
3. Mitigate the difference.

The problem in this case was mitigation, because to attenuate the input one required spring and damping elements and these required a finite deflection if the isolator was to meet the first two requirements.

The overall package was extremely tight. After several weeks of pursuing various mechanical arrangements we came up with a design that just might fit. The main vertical isolator was indeed the same one I had used years before for the Minuteman IV. We applied the general rule that nuclear ground motions always initiate with the ground dropping away, so static forces from the isolator would support the missile at almost full compression. Thus, the 1972 vintage Minuteman IV isolator, 0·3m diameter, 1m stroke, and 1800 kN force, was to live again in a different launcher. It would, however, become a pure damper, since it could support the missile weight on an internal compression hard stop.

For the lateral isolators we utilized a dramatically up-scaled version of Taylor Devices' hard-centering isolators for shipboard systems, along with several added semi-active elements unique to the MX program. But there was much more to the complete isolation system than just Taylor's hardware. Martin Marietta suggested that we team with an aerospace firm that could package the isolators, support structures, and control systems into modular shock isolation units, or SIUs.

This was contractually arranged by teaming with a local aerospace firm, the Bell Aerospace Company. The Bell- Taylor team was eventually awarded a $60 million development contract for the MX-SIU (Figure 19). All of us thought we were finally on a program 'that we can retire on,' this being the ultimate scenario in the aerospace/defense field. Unfortunately this was not to be, for Jimmy Carter lost the Presidency to Ronald Reagan, who promptly came up with his own basing mode scheme. Eventually the MXSIU program died the same slow death as the previous Minuteman IV. But the vertical isolator that dated to the old Minuteman IV would live yet again.

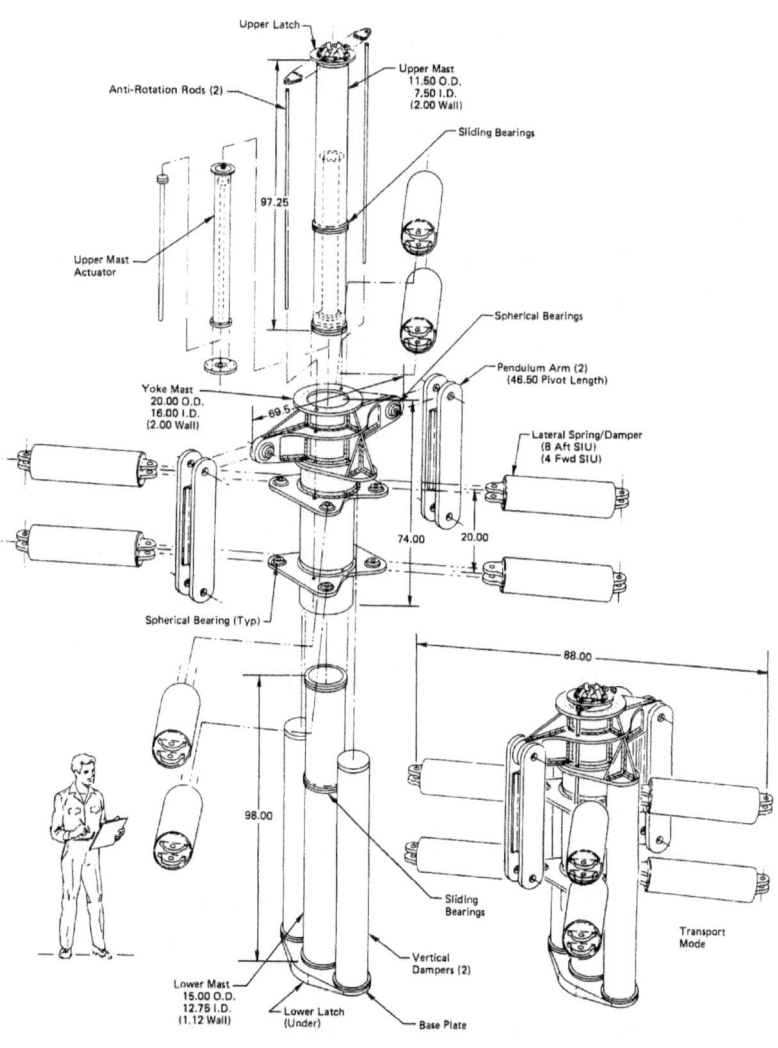

Figure 19. MX-SIU. Photo Courtesy Textron Defense

9. THE REAGAN YEARS

The first inkling Taylor Devices and Bell Aerospace had of the new basing mode was from a press release by Casper Weinberger, Secretary of Defense, that described a launch system called 'Dense Pack,' where all the MX missiles were to be clustered at just a few visible sites, launching from a vertical silo. When the enemy attacked, Weinberger claimed they would launch all missiles at once, with each MX launch site being attacked by many weapons. He stated that all the exploding warheads would cause 'fratricide,' where the warheads would blow up each other before actually hitting the launch site.

Engineers on the MX program informally renamed the program 'Dunce Pack' because it assumed our enemies were completely stupid. All of us in the field knew that to attack a Dense Pack site, you would only need to lob in warheads one at a time, spaced a few minutes apart in a so-called 'pin-down' attack. The MX would not have a chance to launch, and the detonating warheads would eventually carve out a huge crater with the MX launchers tipped over and useless. Bell decided to dropout of the program and Taylor Devices eventually teamed with Boeing Aerospace to build the Dense Pack shock isolators. These isolators ended up having more than 8m of stroke, far beyond Taylor Devices' manufacturing capabilities. Boeing said they could handle the large parts, with Taylor Devices building the seals, piston heads and orifice configuration. Fabrication was started on one isolator before the game changed- Dense Pack died- replaced by studies for a new 'optimized" solution.

In response, Taylor Devices submitted a proposal to the US Air Force to perform basing mode studies on the MX. We centered our proposal upon writing a 'quick look' structural software code that could handle virtually any type of isolator parameter, but limited the structural model of the missile and structure to just a few lumped masses or beam elements. The reason for this imposed limitation was simply that if the isolator was working correctly attenuation of the inputs would be such that the isolated mass and the launch building or structure would stay well below yield at all points, thus negating the need for elegant modeling of these elements. The only available validated software at that time which could adequately model highly nonlinear isolators and high-speed transients was MSC.Nastran®, a highly complex code requiring long set-up and run times. Our goal was far different: a simple code that could run on a PC- a PC of 1980s capabilities.

Our proposal was accepted and David Lee and I embarked on a voyage into the software business. With the aid of some young engineers hired from U/B, we accomplished this quickly, borrowing heavily from several 1960s era analog-based codes long forgotten by most engineers. The resulting code could handle complex isolator parameters, including:

1. variable damping constants based on position or velocity;
2. any damping exponent from 0 to 2;
3. damping roll-off or roll-on at certain frequencies;
4. mechanical damping filters or disconnects (gapping or detents);
5. time-sensitive damping;
6. nonlinear springs;
7. gapped or detented springs, either linear or nonlinear;
8. input time steps down to 0·0001/seconds;
9. precursor inputs/signals.

The Air Force liked this software program very much, and began sending numerous new problems to us, which we realized were often related to new MX basing modes that we had heard about in the media. The Air Force was also intrigued with the way some of the isolator parameters could be varied. In fact, many of the methods we used to make the hardware perform in specific ways had never been disclosed outside of the specific project they were originally developed for.

None of this damping and isolator information and technology had ever been disclosed to the public at this point because of security restrictions. In more than a few cases, after we disclosed the information or simulation results to our military customer, it would be classified to a level so high that Taylor Devices' employees, even though they were the originators of the data, did not have a high enough security clearance ever to see the data again! In any event, the Air Force did share the information with their major contractors, and the company subsequently received two lucrative development contracts for dampers on the US Air Force B-2 Stealth Bomber and the US Navy Sea Lance Missile.

The 1980s were looking good; the company added engineers to handle all the software and development work, and we added production people to build all the new military hardware that we added to our commercial lines. But even more changes were on the horizon, initiated by U/B and the State of New York.

10. SEISMIC DAMPERS?

In the mid 1980s corporate changes occurred at Taylor Devices. Paul Taylor, now in his late sixties, was having health problems. I had been promoted to Executive Vice President when P.T. went into the hospital for cardiac bypass surgery and what would be a 4-year recovery period from 1983 to 1987. When he returned to work (he always said he would never retire), he expressed concern that the company was doing too much military work and should attempt to boost commercial sales. The problem was that while government sector sales were booming, the traditional crane buffer sales were flat- the steel mills were consolidating and closing.

At this time, an NY State Assemblyman from our area, Robin Schimminger, had just started a new program of small business grants from the NY State Office of Science and Technology. Assemblyman Schimminger suggested that Taylor Devices apply for a grant to attempt to commercialize our missile isolator technology and our software. After thinking about commercial topics for a while, all I could come up with was the 'crazy idea' of putting dampers and isolators into buildings for seismic protection. However, the program offered a pure grant of up to $25000 for studies and reports, so I wrote a proposal and submitted it. A few weeks later we indeed received the grant, and I assigned two of Taylor Devices' engineers to alter our software to look at buildings and earthquakes. We knew little about earthquakes or conventional buildings, but applied what we did know about MX missile launch structures. Our study provided two possible uses for the technology, either as part of a base isolation system or providing distributed damping within the building frame.

David Lee was tasked with going over to Caltech, his old alma mater, to get some seismic transient records. When we reviewed these records we were concerned that our dampers and isolators might be overkill; the seismic transients were peaking at only 0·5m/s rather than the 10m/s and higher speeds of nuclear ground motions. When we finished our analysis, we discovered that the same low-exponent damping that we used for the Air Force also worked well if our building structure was taken into the inelastic range. If the building was to remain elastic, then a purely linear damper seemed to be optimal. In our report we assessed a hypothetical 6- to 10-story building and found that about a dozen small (at least to Taylor Devices) 75-ton force linear fluid dampers within the building frame in each direction could at least double the velocity of shaking that the structure could withstand before it yielded. The dampers produced similar beneficial results whether used within the frame or as part of a base isolation system with relatively low or 'flat' rate springs. We did not discuss the post-yield condition simply because it was classified for us to disclose that low-exponent dampers even existed.

After we submitted the report we were contacted by Professor Larry Soong from U/B, who was heading up a new program at the university called NCEER, the National Center for Earthquake Engineering Research. Over dinner we discussed various items and it became evident that NCEER was going to be looking at active seismic control schemes. It was also obvious that Larry wanted me to donate to NCEER's research. I told him that Taylor Devices had thoroughly evaluated active control for large structures for the government, and that the amount of power required and its reliability during a seismic event made it non-feasible. I told him that Taylor Devices was committed to passive control devices for large structures, and that we would not participate. We also had a more pressing issue: by1988 the winds of change were blowing for the defense industry; it was becoming apparent that the Cold War was coming to an end. After decades of concern about the Soviet threat, the Cold War appeared over and that meant trouble for future contracts. We did not realize it at the time, but in reality our relationship with NCEER was just beginning.

DAMPERS: HOW VISCOUS DAMPERS PROTECT STRUCTURES

In late 1988, David Lee and I were at a periodic program review for our missile basing contracts and were informed that since the Cold War was ending our contracts would not be renewed. That evening the two of us had dinner, and sat sketching on our napkins what we thought could be new uses for our missile isolators. We came up with two items. David sketched an MX missile in a horizontal position, then added a propeller to replace the rocket nozzles and a conning tower on one side. He said, 'I've now made the MX into a submarine- a new market.' In turn, I sketched the MX in a vertical position, squared off the warhead fairing, drew a ground line near the engines, and added windows. I said, 'I've now made a building out of the MX, and it has seismic and wind protection from our dampers.'

In the end, both of these sketches proved prophetic. Taylor Devices went on to provide versions of our former MX isolators for the US Navy's Seawolf and Virginia Class submarines. And I went back to talk to the people at NCEER, after obtaining permission from the Air Force to discuss advanced damping concepts and components publicly. After all, if the MX had no more enemies to fight, then perhaps the isolation components could go on to commercial use fighting earthquakes.

11. NCEER AND THE SAN BERNARDINO MEDICAL CENTER REPLACEMENT PROJECT

When I returned to NCEER in 1989, the center was immersed in its planned research projects. Some of the instructors I had in the 1960s were still there, but the research programs were now dominated by a new generation of PhDs. Since the 1989 Loma Prieta quake in the San Francisco area had just occurred, they were interested in new and larger seismic inputs. A meeting was arranged with Dr. Michael Constantinou (now Professor Constantinou) to discuss damping applications. As we began our discussions it was obvious that we both had strong opinions about structural control that were somewhat different. After a while I realized that Michael's experience included friction, viscoelastic and active/semi-active elements, but large fluid dampers were evidently not well understood outside the military. We both agreed that many problems occurred from engineers just assuming that all structures had around 5% damping, this being extremely optimistic. Michael mentioned that one of his graduate students was doing testing on a scaled steel moment frame of 1500kg weight, and that actual tested structural damping was only about 2% critical when the frame was loaded to the level where inelastic motion was first evident. He commented that he would like to add fluid dampers until the damping was 5% total. I laughed and commented that if the input excited the first mode of the 5%damped frame and lasted for long enough to achieve full mechanical resonance, then all this would do is cut the amplification from 25 to 1 to 10 to 1. The building would still collapse in either case. I suggested going to 25% damping as a first step, cutting potential resonant amplification to only 2 to 1. Michael inhaled deeply and said, 'But the dampers will be huge.' I did a quick calculation and replied, 'No, they only really need to be 2 units of 25mm diameter and about 1kg weight, and if I use a safety factor for prototyping the unit it will go up to 38mm and 1·5kg.'

This surprised Michael and he presented a second issue: 'But the damper will have very high force. What if it excites a higher mode in the test structure?' To this I replied, 'No problem. I'll rig a linear damper with an automatic passive roll-off of its response above 5 Hz or so…' Michael cut me off, 'You can do all this with a fluid damper?' I replied that Taylor Devices' dampers could indeed do this, and that I would deliver prototypes in 4 weeks. I did this by 'borrowing' some units off our production line for the B-2 Stealth Bomber, removing the government part numbers and making some minor changes.

These prototype dampers made amazing improvements in the test frame performance. Without dampers, the bare frame was at the onset of yield at only one-third the level of the El Centro earthquake. By adding just two small dampers on diagonal braces, the frame could handle the entire 100% level El Centro quake without increasing base shear loads or drift (Figures 20 and 21). The simple facts were that fluid dampers inherently provide a response that is completely out of phase with the normal flexural stresses in the frame. All other types of dampers that Michael had tested previously had an essentially in-phase response, thus cutting deflection while raising the stress.

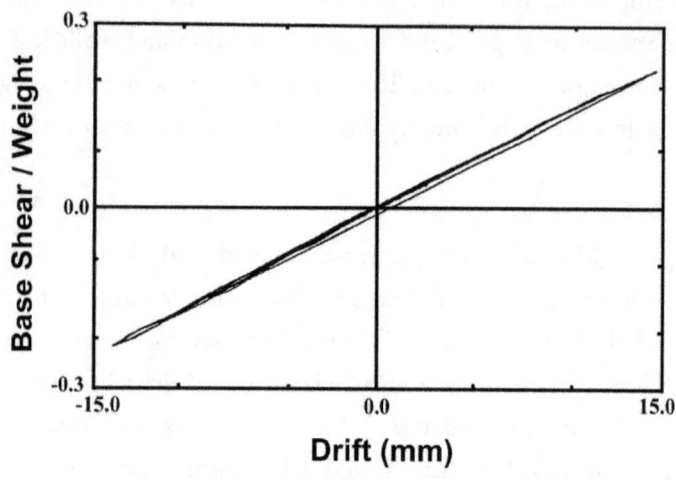

**1-Story, No Dampers, El Centro 33.33%
Total Damping = 2%**

Figure 20. Steel Moment Frame Building, No Dampers

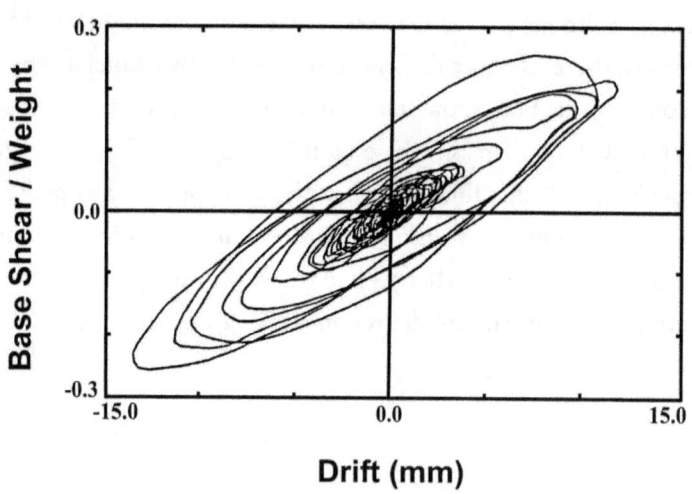

**1-Story, 2 Dampers, El Centro 100%
Total Damping = 22%**

Figure 21. Steel Moment Frame Building, With Dampers

In addition to the basic fluid viscous damper, Taylor Devices also provided NCEER with some complete isolation units combining liquid springs and dampers. We even built a damper equipped with a functional semi-active control system for testing by NCEER Director Professor Masanobu Shinozuka.

All of these prototype devices led to a series of NCEER Technical Reports, the seminal report being Constantinou and Syman's NCEER-92-0032 (Constantinou and Symans, 1992), documenting the improvements from the use of added fluid damping in a steel moment frame. Additional reports were published over the next few years (Lambrou and Constantinou, 1994; Tsopelas and Constantinou, 1994). The only problem encountered was when Michael needed to know where the test dampers came from for inclusion in the report. As mentioned, they were from the top-secret B-2 Stealth Bomber project, so I was required to ask permission to disclose the project's name. Expecting the worst, I made a few phone calls. After my contacts revived from the initial shock, they eventually replied 'Okay, you can tell the world that parts from the B-2 Stealth Bomber will protect buildings from earthquakes. Just make sure you give credit to Northrop, Boeing, the United States Air Force, and the Department of Defense.' How times had changed . . .

By mid 1991 other things had changed, and I became President of Taylor Devices. Dad had never fully recovered from his bypass surgery and the company's Board of Directors decided it was time for him to retire. His retirement did not come easily and negotiations took place over the balance of 1991 that ended fairly for both Taylor Devices and its founder. Paul Taylor's company stock appreciated nearly tenfold after he retired, leaving him money and time for all the things he was too busy to study and learn about during his professional years. He passed away a year after his wife Isabel in 2004.

In 1993 NCEER debuted the fluid damper test results at the ATC 17 Seminar in San Francisco. Michael Constantinou had warned me not to expect much, since he thought that it would take at least 5 years before anyone would consider such radical technology. But Michael was wrong. Our first little poster display drew a continuous crowd and we resorted to having the hotel run copies of our very brief test results and a one-page product data sheet I had made up on a typewriter, complete with taped-in copies of photos. When I returned to the factory, I was convinced that seismic dampers were poised to become a major product line of Taylor Devices. Since this was such a different market from the company's prior experience, I decided that we needed some sales and marketing help- and quickly. I turned to U/B for help and in mid 1993 Taylor Devices hired Craig Winters, with a recent MS in Civil Engineering, and Bob Schneider, with a BS, also in Civil Engineering. Since they were from U/B, both were somewhat familiar with the concept of added damping in structures. Within a few short months they were busy fielding seismic and wind-damping inquiries, and they still are doing this today as the backbone of our seismic marketing program.

In retrospect, the most important thing that happened at ATC 17 was my opportunity to meet and have extensive discussions with the late Dr Roger Scholl. Roger was the inventor of a hysteretic damper using yielding steel which he called an ADAS element (for added damping and stiffness). Roger also seemed to know just about every engineer at the conference. Our conversation started with me asking him metallurgical questions from the floor when he finished a presentation on the ADAS element. We talked at length and very promptly became friends. The next evening, Roger came to me and said he would like to introduce me to some people who potentially needed dampers for a hospital project called the San Bernardino Medical Center Replacement Project. This would prove to be Taylor Devices' first contract for seismic dampers.

11.1 The San Bernardino Medical Center Replacement Project

David Lee and I already were somewhat aware of this project from a visit we had made the previous month to KPFF Engineers in Los Angeles. During this call, Jeff Asher and Saif Hussain had brought in Professor Gary Hart and people from the Hart Consulting Group to KPFF to discuss the addition of dampers to five buildings of a base-isolated hospital complex. The people Roger introduced me to at ATC 17 were from the project's peer review group and the California Office of Statewide Health Planning and Development (OSHPD). They briefly described the project and the fact that its near-fault location required very long displacement rubber isolation bearings. Design and cost issues included the bearings themselves and the long displacement's effects on utility connections and the parking lot and walkway expansion joints, among other items.

They asked very bluntly if dampers could help and I replied 'Yes.' I stated that the biggest immediate issue was getting the software codes being used for the analysis to accept nonlinear dampers. Using hand calculations, David and I had estimated that 200 dampers were needed, each of 100 tons output for the specified input's peak translational velocity. We knew that Gary Hart had contacted NCEER, and that Michael Constantinou and Professor Andrei Reinhorn were working on the software issue back at U/B. Roger's contacts at ATC 17 stated that if the formal damping analysis cut the deflection enough they would possibly use dampers for the project and put them out on bid.

Two weeks later, David and I were called to Sacramento for a meeting of the peer review committee, chaired by Professor James Kelly of UC Berkeley. The analysis by KPFF and Hart Consulting demonstrated that adding 30- 35% fluid damping to the base isolation system cut displacements by more than 50%. A damping exponent in the 0.4 range appeared optimum. Approximately 20 people were present at the meeting, which reminded me of a typical defense program design review. Thus, I was fully prepared for the 2 hours of rapid-fire questions that followed. The biggest issues were the newness of the technology, quality assurance, and how to test full-scaled hardware. In the words of one attendee, 'OSHPD needs new technology to solve this problem, but the new technology has to have been in use for at least 10 years to be considered!' I was fully prepared for this, and indeed the unit we were proposing met these requirements: it was 0·3m diameter and had 1m stroke with 1800kN maximum output force- the very same device used for the early 1970s Minuteman IV, and the late 1970s MX-SIU missile launchers. I also had drawings and photos of finished hardware with me to support this. Quality assurance proved a non-issue, since we explained that all Taylor Devices' products were manufactured under formal US Defense Department standards subject to periodic review and audit.

The question of testing was also resolved. The peer review committee desired full-scale cyclic testing- but no such test machine existed for this large a damper. In fact, the only existing test facility to dynamically test at 1800kN at full seismic speeds was Taylor Devices' large drop test machine. Professor Kelly suggested having NCEER test a 1/6 scale model on their hydraulic cyclic test machine. This would be followed by correlating the scale cyclic test data to testing of the same unit on Taylor Devices' vertical drop tower. The full-sized units would then be tested only on the drop tower, with multiple drops at various velocities and positions, both in compression and extension. This testing plan was found to be acceptable by the peer review group.

By 1994, Taylor Devices had received a contract for 186 dampers of 150 tons output for this project, which formally was named the Arrowhead Regional Medical Center (Figures 22- 24). This was followed by subsequent orders for the very same damper for the retrofit of the Los Angeles City Hall. Over the years, we have manufactured this same damper for numerous other base isolation projects. Our production of this unit is now many times the quantity ever built for its previous use on the Minuteman IV and the MX. If one had to name a quintessential example of commercialization of defense technology, then I would submit this product for consideration. As the old English saying goes, it is quite literally a conversion from 'swords to plowshares.'

Testing of the San Bernardino Dampers required thousands of separate drop tests on our old Apollo Program drop tower. Unfortunately, towards the end of the production tests, we noticed a shift in the ground node frequency. This was very bad, since it indicated that the shale bedrock serving as part of the ground node was failing. When we began the San Bernardino tests, the measured ground node frequency was 165Hz, compared with 170Hz when the rig was new, some 30 years previous. After the testing of the last production San Bernardino unit, the nodal frequency was down to 100Hz. We elected to abandon the test rig shortly after, and took out a construction loan to build a new facility. This time, we stayed away from the bedrock and poured a 1500 ton reinforced concrete block mass for the ground node, set well below grade. Hopefully this facility, rated for up to 3000 tons damper force with a 900Hz ground node frequency, will last for many decades. Subsequently, the company internally funded a series of high-speed hydraulic test stands capable of cycle testing large dampers having up to 1000 tons of output force.

Figure 22. Artist's Rendition of the San Bernardino Medical Center

Figure 23. Doug Taylor with the First San Bernardino Damper

Figure 24. San Bernardino Medical Center Complex
(Arrowhead Regional Medical Center)

12. OTHER EARLY SEISMIC DAMPER PROJECTS

12.1 Pacific Bell

While the Arrowhead dampers were being built, the 1994 Northridge Earthquake occurred, elevating interest in dampers for all types of structures. We were contacted shortly after the quake by Gregg Haskell, a structural engineer from Cole, Yee, Schubert who was working on the three-story Pacific Bell North Area Operation Center in Sacramento. This steel-frame building was under construction when the Northridge quake occurred and Pacific Bell had asked that Gregg and Tom Hale, also of Cole, Yee, Schubert, rerun the analysis for their structure using the Northridge transients- even though the building was not located in a hazard zone as severe as the Los Angeles area. Greg discovered that the severe Northridge inputs required either substantial modifications to the building or the addition of seismic dampers. Gregg and Tom were also concerned about using hardware that had not under-gone extensive testing. Luckily this building used relatively small dampers in a unique bracing element. The only other damper we had available other than the big base isolation unit was the 1/6scale prototype that NCEER had tested for the San Bernardino project. We proposed this unit to Gregg and Pacific Bell, and they elected to add 62 pieces of this damper to the structure, installed in a bracing design subsequently patented by Gregg as the 'Velocity Brace.' When completed, this building was the first structure to be operational with fluid dampers for seismic protection (Figures 25 and 26).

Figure 25. Pacific Bell North Area Operation Center – Sacramento, CA

Figure 26. Roger Scholl with Finished Pacific Bell Installation

12.2 The Woodland Hotel

Shortly after starting on the Pacific Bell dampers we were contacted by Kit Miyamoto, President of Miyamoto International, Structural Engineers, who was working on the retrofit of the four-story Hotel Woodland in Woodland, California. This historic concrete structure had to be retrofitted in such a way that its outer construction and appearance were preserved. This was so the building could retain its historic status. This proved to be a challenging project, and we asked for help from Roger Scholl. Roger's aid proved invaluable, and in short order Kit was able to design a retrofit that preserved both the exterior appearance and the interior room size. A total of 16 dampers of 50 tons output force were needed. A unique arrangement of steel beams was used to 'box' the bays having dampers with the units installed in chevron bracing to preserve the door entranceways (Figures 27 and 28).

Figure 27. Woodland Hotel

Figure 28. Roger Scholl with Woodland Hotel Dampers at Installation

12.3 The Money Store National Headquarters

Kit Miyamoto had become a true believer in the ability of added damping to reduce seismic stress and deflection. The new-build Money Store project challenged both Kit and Roger due to its unique structural considerations (Figure 29). This structure was 11 stories in height, with a unique pyramid shape, causing a very diverse set of structural modes (the owners' desire was to have a building that looked like an Egyptian pyramid). Kit found out that since fluid damping suppresses the response of all modes, putting in a fairly large number of dampers was a good solution. His design used a total of 120 dampers rated for forces of 130 tons and 70 tons. To maximize entrance space to the offices and window area, Kit took advantage of the heavy 'post-Northridge' moment frame, and installed the dampers in short diagonal braces meeting at the center of the bays (Figure 30). This also served to reduce costs, since each of the short diagonals would have substantially greater column strength for a given extender cross-section than a single long diagonal.

Figure 29. The Money Store

Figure 30. Craig Winters with the Money Store Dampers

In just a short time, David Lee and Roger Scholl were splitting the State of California for sales visits- Roger in the north and David in the south. David decided that additional market coverage was needed at the owner and contractor levels, so we discussed the addition of a second layer of representation. Towards this goal we added McQuarrie Associates in the San Francisco area and International Marketing Concepts in the Los Angeles area. Unfortunately the multiple layers and north-south division did not work out as planned due to the sudden untimely death of Roger Scholl. As an end result, David Lee elected to work at the engineering level throughout the Western USA and to have regional representation established at the owner/contractor level.

Meanwhile, in the Eastern USA, Taylor Devices' representatives were also looking for damper applications, although we assumed that interests in this area would be more related to wind. Paul Tuttobene, the company's representative in the New England area, was particularly skeptical of getting much interest in his territory. One day Paul and I were calling on Raytheon in Boston and finished our last call at 2 p.m., with an 8 p.m. return flight to Buffalo. I had brought along some inquiries for dampers from Boston, so we tried a cold call on Bob McNamara of McNamara/Salvia, Inc., who had mailed in an inquiry. Bob's interest in dampers was strong, especially since he had previously worked on tuned mass damper applications on office towers for wind motion suppression. Bob was also a UC Berkeley graduate, so he was very much aware of seismic design requirements. We ended up talking to Bob until 5 p.m. and left with the feeling that we would meet him again soon.

In the next few months, Paul became very active, calling on structural engineers in the Boston and New York City areas. As a result, Taylor Devices received orders for our first two large office tower projects: the 28 State Street Building in Boston, and Torre Mayor in Mexico City.

12.4 28 State Street

The 44-story 28 State Street Building was constructed in 1969, but had been taken over by the US Government (Figure 31). The building had a history of uncomfortable motions during wind storms and Weidlinger Associates of Boston was given the project of improving occupant comfort as part of a complete renovation of the tower. The original analysis of the structure took place in the pre-computer age, and Weidlinger's new studies indicated that a torsional mode combined with the in-plane motion were generating responses that were potentially uncomfortable to the occupants. Using dampers located as close as possible to the building's outer perimeter allowed suppression of both modes. As is typical of most buildings of 30 stories and up, damping levels in the 10% range are usually sufficient to provide wind motion suppression, good seismic performance, and occupant comfort. As a result, only 40 dampers were needed to provide the desired response, each rated at 70 tons of output force (Figure 32). Of more importance to Taylor Devices was the fairly short maximum deflection in these dampers, at only plus or minus 22mm. In most cases, interstory drift in this structure was small enough such that only 25% of the available damper stroke was utilized.

Figure 31. 28 State Street – Boston, MA

The results verified that tall buildings appeared to require large-diameter dampers with a short stroke, and this is not a very cost-effective solution due to manufacturing costs. In terms of general rules, the most cost-effective damper for use in a bracing system is one with a maximum total stroke roughly the same as its maximum diameter. In the case of 28 State Street, the damper OD was 225mm, with only a 50mm total stroke. I pondered this problem at length and finally decided to have a long conversation with Roger Scholl on this item.

Figure 32. Craig Winters and Paul Tuttobene with 28 State Street Dampers at Installation

13. DAMPING CONCEPTS FOR TALL BUILDINGS

Roger and I had spent hours in conversation discussing the best way to apply dampers into tall buildings. The first thing we agreed on was that if the analysis indicated that simple diagonal or chevron braces could be used with a total damper displacement of plus or minus 25mm (including combined loadings, thermal displacement and adjustment displacement), then this was probably a reasonably cost-effective solution. An alternative was to use a single damper and extender to span several floors, thus combining the deflections of each of the spanned floors into that for the single damper. The down-side of this concept was the need to support the brace extenders to prevent buckling. If more than two floors were spanned, an unsupported extender brace became massive due to column strength considerations.

A second alternative was the solution of mechanical engineering: add a mechanism to reduce the damper force and increase its deflection simultaneously, holding the energy dissipation per cycle constant. Roger stated that this had been looked at previously in Japan with rotating levers about a fulcrum driving the damper. However, he explained that mechanisms like this normally will not work in a building because they either bind or buckle when exposed to out-of-plane motions. I agreed with Roger's assessment, but decided to think more about it in my spare time, recalling the quote attributed to Thomas Edison, 'Invention is 1% inspiration and 99% perspiration.'

As luck would have it, I was involved at the same time on a project developing concepts for a recoil reduction system for shoulder-fired weapons. I had gone to the public library and checked out a selection of books on firearms history, a field where elegant mechanisms have been used for hundreds of years. One night I was reading about arms of the late 1800s and came across a cutaway from George Luger's patent of 1900, using a so-called toggle mechanism to lock and unlock the breech of a pistol (Figure 33). This pistol later became the famed Parabellum, or so-called 'German Luger' pistol of the early and mid 1900s.

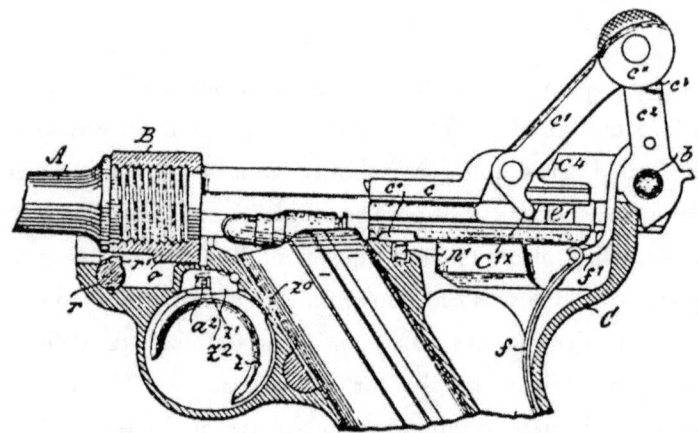

Figure 33. Toggle Breech Mechanism – Model PO8 Luger Pistol

I was amazed at the simplicity of this mechanism and its inherent reliability. I also noted that the mechanism was extremely stable against out-of-plane motion. The next day I had grafted a version of this mechanism into a sketch of a building bay, and called Roger. After a long (and somewhat heated) discussion, he agreed that the toggle I had sketched seemed to address the problem, and that it would indeed accept out-of-plane motions. Alas, Roger did not live to see the toggle brace system I eventually patented which had numerous applications on buildings.

14. TORRE MAYOR

Shortly after Bob Schneider and Craig Winters started at Taylor Devices, Craig took a call from one of his former classmates at U/B who was working at the Cantor Seinuk Group in New York City on the design of a 50-plus-story office tower in Mexico City, an area noted for large earthquakes, soft soil, and very long period motions.

Craig and Paul Tuttobene went to New York and learned more about the planned building and itsmany design challenges. The owner, Paul Reichmann of the famed Reichmann family of Canada, wanted to have the tallest building in Mexico City. He wanted 50 stories plus to dominate the other buildings in the city, all of which were less than 40 stories in height. Cantor Seinuk, working closely with Enrique Martinez Romero of EMR Engineers of Mexico City, knew that in Mexico City the softsoil essentially limits the height of buildings. The local joke was that in Mexico City it is easy to find bedrock, you just dig down 1·5km and you will find it, since much of Mexico City is built on top of the bowl of a long-extinct volcano. Because of the soft soil, the local design codes were such that a conventional building on a conventional site is essentially limited to 40 stories in height by its weight. This was indeed the case with Torre Mayor, and Cantor Seinuk asked how they could use dampers to help. Our suggestion was to run the analysis again, with 15% damping added as a simple macro fora quick look assessment. Cantor also brought in Professor Andrei Reinhorn from U/B to help with the structural analysis and design concepts. The report came back that added damping reduced stress and deflection to well below the allowable limits. My reply was that the next step would be to formally put dampers discretely into the model, adding only 4-5% additional damping, and determining the resulting stress reduction. Then, remove steel from the structure until stresses came back up, and use the steel that was saved to add more floors to the building. The next step was a second iteration, adding some more damping and repeating the process. This is vastly oversimplified, but was indeed the approach that allowed the building with a total damping of 10% to be 57 stories in height (Taylor, 2003). This was a fresh approach to using dampers, and Enrique Martinez Romero was tasked with presenting the analysis and design to the authorities in Mexico City, subsequently obtaining full approval (Figure 34).

Torre Mayor was a very slow-moving project due to many elements outside the realm of engineering and construction. Over the years the project was progressing, Mexico suffered both political and monetary system upheavals. When construction finally began, Ahmad Rahimian of the Cantor Seinuk Group had arrived at a solution with a series of mega-brace dampers on the long walls of the building. Only 28 dampers (14 per wall) were required on the long walls for the entire 57-story structure. The dampers were large, each rated at 570 tons of force, with a stroke of plus or minus 52mm. Each damper spanned multiple floors. The short walls of the building used 74 pieces of a smaller damper rated at 280 tons force. These also spanned multiple floors (Figures 35 and 36).

When the building was completed, it began experiencing earthquakes at a rate averaging one per month. In January 2003, a magnitude 7·6 earthquake generated strong shaking in Mexico City. Torre Mayor not only was undamaged, but many occupants reported that they had no idea an earthquake had actually occurred until they heard sirens outside and saw people gathering in the streets.

In 2003, Torre Mayor was a finalist for the Charles J. Pankow Award for Innovation from the ASCE's Civil Engineering Research Foundation. In 2005, Torre Mayor won the Pankow Award for its innovative bracing and damping systems.

Figure 34. Artist's Rendition of Torre Mayor

Figure 35. Torre Mayor

Figure 36. Enrique Martinez Villalba with Torre Mayor Dampers

15. TALL BUILDINGS WITH TOGGLE BRACES

Torre Mayor demonstrated that spanning multiple floors with dampers in tall buildings is cost-effective, yet allows the architect to have substantial design freedom. The patented toggle brace concept was an alternative approach, using a mechanism to multiply small interstory drift to a level where a damper can also be very cost-effective. I received four patents on the toggle brace damper over the period 1999- 2002, and testing at U/B went exceptionally well (Taylor and Constantinou, 1998; Constantinou et al., 1997). A variant on this concept, the scissor brace damper, was later patented by Professor Michael Constantinou. The toggle braces have proved highly successful, with three tall buildings installing toggle brace dampers in the year 2000. These include McNamara/Salvia's 38story 111 Huntington Tower in Boston, and DeSimone Engineers' 37-story Millennium Place Tower in Boston and the 37-story Yerba Buena Tower in San Francisco. Thus, between the technology of mega braces and toggle braces, Taylor Devices now has damper arrangements that will optimize nearly any structure above 40 stories in height, with conventional bracing methods available for those that are less tall (Figures 37 and 38).

Figure 37. Yerba Buena Tower

Figure 38. Bob Schneider with Toggle Brace Damper at Yerba Buena Tower

16. THE FUTURE

In 1998, the NCEER program ended and a new national engineering research center was founded at U/B, called MCEER (Multi-Disciplinary Center for Earthquake Engineering Research). The center's goal is to evaluate and find improved means for providing resilience to communities subjected to the adverse effects of natural hazards. I am proud to be a member of the MCEER Industrial Advisory Board.

Dampers are a primary means to obtain structural resilience, and MCEER's activities have generated many new applications. For example, when MCEER was founded, a total of 32 buildings and bridges were using Taylor Devices' dampers. Today, more than 700 structures use this technology for protection against multiple hazards. These include:

- Earthquakes
- Hurricanes
- Tsunamis
- Homeland security issues
- Wind-induced vibrations
- Traffic- and crowd-induced vibrations

Similarly, in 1998 there were essentially three ways to install fluid dampers into a structure: either as part of a base isolation system, as diagonal brace elements, or as chevron brace elements. Today, there are ten new ways to utilize dampers, all resulting from MCEER's research. These include:

- Base isolation with spring damper elements
- Base isolation with detented lock dampers
- Base isolation with visco-elastic spring dampers
- Toggle brace dampers
- Scissors brace dampers
- Fluid visco-elastic damper bracing
- Short-radius mounted bridge cable dampers
- Modular wall dampers for residences
- Low-exponent dampers
- 'In situ' retrofit concepts

In terms of disseminating the knowledge of MCEER to the profession, one needs only to look at the volumes of reports the center has published and the many students who have graduated to academic and professional positions after their affiliation with MCEER. Today, many professors such as Michael Constantinou, Andrei Reinhorn, and Andrew Whittaker are hard at work at U/B and MCEER educating the engineers of tomorrow, who in turn will have even better knowledge of the uses for damping than those who went before them.

If it were not for the open-minded interest of NCEER and MCEER, from top to bottom in both their people and research approach and attitude, dampers would not be part of today's structural engineering field.

The 700+ structures now using dampers include many special construction projects, including:

- 18 airports
- 28 hospitals
- 16 stadiums
- The Los Angeles and Hayward, CA, City Halls
- The San Francisco Civic Center
- The San Francisco Oakland Bay Bridge

- London's Millennium Bridge
- Sixteen towers, including the Petronas Twin Towers in Malaysia

Indeed, in my opinion the use of dampers in structural applications has really just begun. In the future the use of this technology will continue to expand in scope. As for me, I will always continue to be open to new ideas- wherever they may originate- and I will never be afraid to get my hands dirty making sure that what I have designed actually fits together and works as planned.

REFERENCES

Constantinou M, Symans M. 1992. Experimental and analytical investigation of seismic response of structures with supplemental fluid viscous dampers. *Technical Report NCEER-92-0032.*

Constantinou M, Tsopelas P, Hammel W. 1997. Testing and modeling of an improved damper configuration for stiff structural systems. *Technical Report Submitted to the Center for Industrial Effectiveness and Taylor Devices, Inc.*

Lambrou V, Constantinou M. 1994. Study of seismic isolation systems for computer floors. *Technical Report NCEER-94-0020.*

Taylor DP. 2003. Seismic dampers for the Torre Mayor Project. In *Proceedings of the VII International Symposium for Steel Construction,* Veracruz, Mexico.

Tsopelas P, Constantinou M. 1994. Experimental and analytical study of a system consisting of sliding bearings and fluid restoring force/damping devices. *Technical Report NCEER-94-0014.*

Taylor DP, Constantinou MC. 1998. Development and testing of an improved fluid damper configuration for structures having high rigidity. In *Proceedings of the 69th Shock and Vibration Symposium.*

DOUGLAS P. TAYLOR
President

Mr. Taylor holds a B.S. degree in Mechanical Engineering from the State University of New York at Buffalo, awarded 1971. He has been employed by Taylor Devices, Inc. of North Tonawanda, NY since 1971, and was appointed President in 1991. Mr. Taylor previously was President of Tayco Developments, Inc., an affiliate of Taylor Devices, Inc., where he had been employed since 1966, and was appointed President in 1991. He is inventor or co-inventor of 35 patents in the fields of energy management, hydraulics, and shock isolation.

Mr. Taylor is widely published within the shock and vibration community. He has authored more than 75 technical publications on diverse topics ranging from automotive crash safety to ship and spacecraft survivability.

In 1998, Mr. Taylor was awarded the Franklin and Jefferson Medal for his commercialization of defense technology developed under the U.S. Government's Small Business Innovation Research program. In 1999, Mr. Taylor was awarded the Clifford C. Furnas Memorial Award by the Alumni Association of the University at Buffalo for his accomplishments in the field of engineering. In 2006, Mr. Taylor was awarded the Dean's Award for Engineering Achievement by the School of Engineering and Applied Sciences at the State University of New York at Buffalo. Also in 2006, Mr. Taylor was named Structural Engineer of the Year (2006) by the Engineering Journal, "The Structural Design of Tall and Special Buildings." In 2015, Mr. Taylor received the Moisseiff Award for contributions to the science and art of structural design by the American Society of Civil Engineers (ASCE). In 2015, Mr. Taylor was inducted into the Space Technology Hall of Fame by NASA and the Space Foundation. Mr. Taylor is also a founding member of the International Association on Structural Control and Monitoring, a life member of the Association for Iron and Steel Technology, and has served on the ASCE committees on Seismic Performance of Bridges and Blast Protection of Buildings.

www.ingramcontent.com/pod-product-compliance
Lightning Source LLC
Chambersburg PA
CBHW050217230526
45470CB00001B/423